Willard Cochrane and the American Family Farm

Volume 14 in the Series
Our Sustainable Future

SERIES EDITORS

Cornelia Flora
Iowa State University

Charles A. Francis
University of Nebraska–Lincoln

Paul Olson
University of Nebraska–Lincoln

Richard A. Levins

Willard Cochrane and the American Family Farm

ARCHBISHOP ALEMANY LIBRARY
DOMINICAN UNIVERSITY
SAN RAFAEL, CALIFORNIA 94901

University of Nebraska Press
Lincoln • London

Publication of this book was assisted by a grant
from The Andrew W. Mellon Foundation.
© 2000 by the University of Nebraska Press
Foreword © 2000 by John Kenneth Galbraith
All rights reserved
Manufactured in the United States of America

Library of Congress Cataloging-in-Publication Data
Levins, Richard A.
Willard Cochrane and the American family farm/Richard A. Levins.
p. cm.—(Our sustainable future ; v. 14)
"Selected writings of Willard Cochrane, 1939–1997":p.
ISBN 0-8032-2935-6 (cloth : alk. paper)
1. Cochrane, Willard Wesley, 1914– . 2. Agricultural
economists—United States—Biography.
3. Agriculture and state—United States—History.
4. Family farms—Government policy—United
States—History. 5. Agriculture—Economic
aspects—United States—History. 6. Agricultural
subsidies—United States—History. 7. Farmers—
United States—Economic conditions.
I. Title. II. Series.
HD1771.5.C63 L48 2000
338.1′092—dc21
[B]
99-052954

For Turner Oyloe

Contents

List of Illustrations	viii
Foreword	ix
Acknowledgments	x
1 Family Farms in Form but Not in Spirit	1
2 The Golden Age	13
3 The Treadmill	26
4 Professor Cochrane Goes to Washington	44
5 An Unreconstructed Liberal	62
6 Heartland	79
Notes	83
Selected Writings of Willard Cochrane, 1939–1997	85

Illustrations

Following page 34
Willard Cochrane advising candidate John F. Kennedy, 1960
Cochrane's father, Willard Cochrane Sr., circa 1907
Cochrane with his mother, Clare Cochrane, circa 1928
Cochrane with his bride, Mary, on their wedding day, 1942
Cochrane talking to a Thai farmer, 1948
A meeting with the minister of agriculture of the Philippine government, 1973
Cochrane lecturing at the University of Minnesota, circa 1980

Foreword

John Kenneth Galbraith

In my early life I escaped the oft-described civility and less celebrated labor of a family farm into agricultural economics. I was concerned with agricultural policy; it produced one of my more notable comments from John F. Kennedy when he was president or a presidential candidate. It was, "I don't want to hear about agricultural policy from anybody but you, Ken, and I don't want to hear about it from you either." I was, thereafter, more reticent on this subject, but Kennedy did not escape. The continuing problem of the small or mid-size farm and its survival in a world of great corporate enterprises continued. It could not be wished away and the principal source of recommendation and guidance came through Kennedy's Secretary of Agriculture, the former Minnesota Governor Orville Freeman, and ultimately from Willard Cochrane. This was not his first guidance; nor was it his last. No one else in the last thirty or forty years has been so intelligently influential. And as so often in agricultural matters, this is history that has been sadly neglected. This book fills the large gap. Richard Levins has told of the life and public career of Willard Cochrane, also his academic life, but he has done much more. He has given a vivid account of the context in which Willard Cochrane's thought and policy were relevant. Particular attention is given to the big corporations which surround and invade the farm scene and which, with their greater market power, are the problem of the traditional farmer. All of this is done with care and in good, clear English. I have the greatest pleasure in giving the book a strong recommendation.

Acknowledgments

Although I have done the usual statistical and historical work one might hope for in a book such as this, I have also benefited greatly from my association with Willard Cochrane and a group of his friends. For several years now I have been fortunate to have had lunch at the local Red Lobster every week with Will, Burt Sundquist, Turner Oyloe, Dale Dahl, Vernon Ruttan, and Arvid Knudtson. The influence of their stories, advice, and readings of earlier drafts is everywhere in this book.

Several of Will's other friends from over the years also took time to comment on this manuscript at various stages along the way to publication. Lee Day, Elmer Learn, Martin Abel, Alex McCalla, John Schnittker, Reynold Dahl, and Sherwood Berg are due special thanks. I also took advantage of the time and talents of many of my colleagues and former students: Lynn Hamilton, William Chambers, Michael Duffy, Brian Buhr, John Ikerd, Suzanne Wood, Karl Stauber, Mark Ritchie, Scott Peters, and Carmen Fernholtz come to mind. Patricia Schwarz was a patient and capable typist throughout the production of the manuscript. My editor-in-residence, Jane Dickerson, can take full responsibility for any text that reads well.

My greatest thanks go to Willard Cochrane. He has been both a wonderful friend and a great inspiration to me over the past several years. He graciously offered his time on countless occasions to explain critical parts of the history of American agriculture and the role he played in shaping it. I also had full access to his publications, his correspondence, and his personal papers as I prepared this book. Without Will's patience and dedication to this project, I would not have been able to complete it.

Like all people who take pride in our country, I value certain of our institutions highly: our public school system, our Bill of Rights, sullied as it may be from time to time by overzealous patriots, our near-universal suffrage, and our national forest and park system. All of these, and many others that could be named, give the United States a distinctive flavor; make it, I believe, a good place to live in. But there is one institution I value particularly, one that is currently undergoing rapid change and may be in danger, like the whooping crane, of passing out of existence. It is the family farm—the family farm as it flourished from the Alleghenies to the High Plains and north of the Ohio River. It once provided a way of life as well as a way of business, and to me it provided a good way of life. Now it provides primarily a way of business, and in years to come it may not provide even that in an owner-operator sense. With capital requirements running as high as $100,000 per farm it is difficult to see how these farms can remain family affairs. But what I want to say here is that I think our country will be losing something vital if it loses the institution of the owner-operated family farm.

WILLARD COCHRANE, IN *FARM PRICES: MYTH AND REALITY (1958)*

1

Family Farms in Form but Not in Spirit

Thomas Jefferson was hard at work on his farm when a former president of the French National Assembly visited him in 1796. According to letters the duke later wrote, he happened upon Jefferson "in the midst of the harvest, from which the scorching heat of the sun does not prevent his attendance." He was duly impressed not only with Jefferson the political genius but with Jefferson the lifelong farmer, a man consumed with "activity and perseverance in the management of his farms and buildings." Perhaps the two men discussed government and the excesses of the French Revolution, but more likely Jefferson entertained his guest with stories of crop rotations, sheep raising, and his plow design that would later win high honors from the Agricultural Society of Paris. Only a year earlier, Jefferson had written to another potential visitor that he would "talk with you about [farming] from morning till night, and put you on very short allowance as to political ailment."

We most often think of Thomas Jefferson in stately dress, signing the Declaration of Independence or serving as president of the United States. Seldom, if ever, do we think of the farm that provided his income and sustained his spirit in difficult times: "I return to farming with an ardor which I scarcely knew in my youth, and which has got the better entirely of my love of study."

To Jefferson, farmers as a group were "first in utility and ought to be first in respect"; they were "the chosen people of God if ever He had a chosen people, whose breasts he has made his peculiar deposit for substantial and genuine virtue." Jefferson feared a tyranny not of kings but of the heavy and unrestricted hand of powerful industrialists. Democracy would be best preserved if the new nation's vast and uncharted land would be home to free and equal farmers. A lasting democracy, one that protected the rights of individuals from big business and big government, would be an agrarian democracy.

Jefferson's dream was very much alive at the beginning of the twentieth century. In 1908 President Roosevelt said that "no nation has ever achieved permanent greatness unless this greatness was based on the well-being of the great farmer class, the men who live on the soil; for it is upon their welfare, material and moral, that the welfare of the rest of the nation ultimately rests."[1] Thirty years later, the secretary of agriculture was the main speaker at a celebration of progress in agricultural education. There, before an assembly at Jefferson's beloved University of Virginia campus, Henry A. Wallace chose the occasion to "do reverence to a man who created in considerable measure the foundations on which they (farmers, educators, and government officials) now stand." At one point in his tribute, the secretary stopped to wonder whether Jefferson loved farmers because he thought them essential to democracy or whether he loved democracy because it gave full expression to farmers. Either way, farming and democracy went "hand in hand in Jefferson's mind."[2]

It is difficult to so much as imagine a modern secretary of agriculture speaking philosophically of the family farm as a foundation for democracy. Far more likely, he or she would use the word "free" in reference to global markets, not people, and democracy would take a back seat to efficiency and other such business-minded terms. The professors of agriculture Jefferson fought so hard to establish have drifted as far, or farther, from their heritage. A respected agricultural economist, writing in one of his profession's top journals, recently compared the nation's smaller farms to sports cars, a luxury completely unworthy of public support. We have certainly come a long way since the days of Thomas Jefferson. The total number of farms in the United States has declined from 6.5 million in 1935 to 2 million in 1997. Of the 2 million farms remaining, 1.3 million are classified as part-time, residential, or retirement. Altogether, these 1.3 million farms produce less than 10 percent of total farm sales. On the other extreme, 161,000 farms accounted for almost two-thirds of our farm sales in 1997. Even the largest of these modern farms are losing their independence as enormous corporations that supply their seeds and process their products gain control through various contracting arrangements.

Willard W. Cochrane watched all of this happen from a vantage few can claim. He was born in the autumn of Jeffersonian idealism and saw it in action on his grandfather's farm in Iowa. He became one of the country's premier agricultural economists and carried the standard of liberalism for President Kennedy in the last serious fight to save the family farm. Then, for forty long years, he held to his principles while traditional agriculture faded away into what he once called "family farms in form but not in spirit." This book is about the spirit, not the form, of those farms. That spirit is in part that of political liberalism and in part that of the struggle to maintain economic democracy in a world increasingly dominated by technology and multinational corporations.

As the twentieth century draws to a close, Willard Cochrane hardly seems like he has been retired for almost twenty years. He remains a regular visitor to his University of Minnesota office, where an autographed picture of President Kennedy, signed, "For Dr. Willard Cochrane with esteem and very best wishes," hangs above his desk. In lectures, books, and professional articles he distinguishes himself as what he proudly terms "an unreconstructed liberal." His interpretation of the liberal vision for society was born during the New Deal. It has three important parts, all derived from the Golden Rule of doing unto others as you would have them do unto you. First, the more fortunate members of a society have an obligation to help the less fortunate. Second, this help may be in part provided through private charity, but the government also is obligated to take action. Third, an enlightened government can do a better job than the free market in helping the poor and downtrodden become productive citizens.

His liberal philosophy has always been at the heart of his thinking on farm policy. For Cochrane, farming and its products were fundamentally more important than most other consumer goods. It was one thing for a person to do without, say, a television, and quite another for a person to do without food. Food was more like electricity, education, and the highway system in that it was better regarded as a public utility:

> Where competition has led to ruinously low prices and returns, poor service, or injury to certain persons or groups, government has historically intervened to regularize that competition, to equalize the bargaining power among contending parties, and to redress inequities. (Government was acting in this role when it brought the railroads under the control of the Interstate Commerce Commission, when it gave unions the right to bargain collectively, and when it has tried to provide commercial agriculture with price and income support.) Where the continuous and uninterrupted provision of a product or service was deemed essential to the well-being of the community, government has traditionally granted certain firms the monopoly right to supply the needs of consumers with that product or service, under the supervision of government with respect to such things as rates, safety, and quality: *it has created public utilities*. Now it is pro-

posed here that the government adopt this general policy with respect to agriculture; first of all to ensure producers of reasonably good and stable prices and incomes, and perhaps in some later period, when circumstances require it, to ensure consumers of an adequate food supply at reasonable prices.[3]

The concept of agriculture as a public utility stands in stark contrast to another way economists sometimes use the term "liberal." Trade "liberalization" is a philosophy in which there are very few barriers to international trade and all countries compete in the same global marketplace. Cochrane has never supported the move toward globalization that has now become the predominant theme in U.S. agricultural policy. It was fine to take advantage of export opportunities as they arose, but it was unwise to depend on them year after year. American farmers needed protection from unstable markets, both domestic and foreign, that could best be provided by public programs.

The Kennedy years were the high-water mark for ideas in which agriculture was seen as a public utility. Since then, Cochrane has watched American agriculture struggle under a policy regime that has increasingly relied on free markets and private corporations. The results, to his mind, have been catastrophic. American farmers have been pitted against farmers throughout the world in a contest to be won by whoever could survive at the lowest prices. Initially, the Cheap Grain Olympics seemed like a cakewalk for Team America. It had the most modern equipment, the best seeds and chemicals, and the vast grain storage and shipping network required by an efficient export economy. Unfortunately for U.S. farmers, the very corporations that supported their move to free markets immediately set themselves to the task of taking away the American advantage. They sold the same tractors, the same seed, and the same chemicals throughout the world with great abandon. When the United States embargoed soybean sales as part of its national policy, grain companies began building Brazil into the soybean superpower it is today. Playing one country against another soon became a powerful and profitable strategy. Having lost their technological advantage, American farmers were forced to compete in the same way a person working in an American textile mill competes, that is, by working cheaper.

The free market, whatever advantages it might have, cares not about the size or ownership of the farms that produce the world's cheapest grain. The move toward bigness, toward efficiency, and toward low-paid labor began pushing the "family" out of "American family farm." The general plan of globalization, where farmers around the world produce the same commodities according to the same cutthroat rules, also began taking the "American" out of "American family farm." To complete the picture, it has become fashionable among larger farmers in the United States to refer to themselves as "producers" rather than "farmers."

While most agricultural economists, Cochrane among them, have worried over how to keep farms profitable, Cochrane was among the first to see that there was more to the story. A farm could be profitable in the strict sense even though a larger nonfarm corporation controlled it. Cochrane devoted much of his career to a search for ways of keeping farms independent. This was no easy task, in light of the nature of farm economics, the growth in government, and global corporations that cast long shadows over even the largest of farms. He has been branded a Communist, derided as a "killer of farm freedom," trampled in the stampede to global free-trade policies, and, at long last, vindicated by new types of farming that removed decision making from farmers altogether.

For Cochrane, the Trojan horse of better farming through technology was filled not with bureaucrats who threatened farm freedom but with nonfarm investors who would dictate practices on operations that for all appearances looked like family farms. In the mid-1960s he wrote an uncanny description of the fate of family farms in the year 2000: lacking specific policies to save it, the family farm would likely disappear. A few in his time also might have seen this coming, but only Cochrane saw the real reason why it would happen. In his view, "the problem is not one of physical organization or production efficiency; it seems likely that farms can be organized to make use of the most advanced technology and to require no more labor than could be supplied by an average family plus one hired man." In other words, the then-popular definition of a family farm could, at least in theory, be partially satisfied in the world of high technology he foresaw. One part of the definition, however, could not. Decision making in the family farm world was supposed to happen on the farm, to be by the farmer.

Here lay the real problem. The technology might be put in place on something that looked like a family farm, but no one could afford it with traditional financing methods. Instead, "commercial nonfarm sources: insurance companies, feed companies, processors, and retail organizations" would step in, "and operational control invariably follows financial control. We could thus in 2000 have family farms in form but not in spirit—farms organized into productive units that are supervised and operated by a family with the help of one or two workers, but managed by and with their financial risks assumed by a non-farm organization."[4]

Cochrane further developed his thoughts on the control of farming when he revised his book *The Development of American Agriculture* in 1993. In the second edition, he described the industrialization of farming as the great watershed of late-twentieth-century agriculture. As outside financial interests took control of farming, Cochrane reasoned that profits would no longer go to farmers. Instead, those profits will be "captured by the business firm in financial control." He acknowledged that "independent farmers do not like the idea of becoming paid employees of an outside firm and losing the opportunity of making a profit from their operations." Nonetheless, this was exactly what was happening, and would continue to happen, in what he called the "mature industrial age" of American agriculture.

A case in point is Murphy Family Farms, by far the largest hog producer in the United States. Its principal owner, Wendell Murphy, appeared on the cover of *Forbes* on October 13, 1997, and was estimated to be worth a billion dollars. Wendell Murphy and his family do not, of course, labor tirelessly feeding their 275,000 sows and 6 million market hogs on some giant version of Old MacDonald's Farm. They contract with farmers across the country to do that for them on farms that often look like traditional family farms. *Forbes* rightfully compared the way Murphy does business to fast-food franchising. The article, titled "The Ray Kroc of Pigsties," left little doubt about the role of farmer decision making in Murphy's world: "Murphy tells his farmers exactly how to grow his hogs—when and what to feed them—producing a standardized hog that commands premium prices from big slaughterhouses." This type of farming is called contract farming. A farmer participates in a vertically integrated system by agreeing to raise animals or crops for a larger concern. In some contracts the farmer maintains ownership of plants and animals and sells them under contract to processors; in others, the processor owns the plants and animals and pays the farmer to raise them. Regardless of the exact contract type, a farmer wishing to use practices not specified by the larger company would risk financial ruin. About one-third of U.S. crops and livestock were grown or sold under contract in 1997. The nation's largest farms accounted for more than three-fourths of the $60 billion value of contract production.

In his discussion of the industrialization of agriculture, Cochrane also recognized a newcomer on the scene, biotechnology. He was at the time unsure of how successful biotechnology would be. He was sure, however, that any successful applications would be "incorporated into the industrialization process." Since he made that prediction, biotechnology has brought about the ultimate strategy for controlling farming. Companies with sufficient research and development capabilities can now use patent laws to own the living organisms necessary for food production. Genetically modified soybean plantings went from 1 million acres in 1996 to 20 million acres in 1998; figures for corn and cotton during the same period are equally dramatic. The goal of all this was spelled out nicely by an agribusiness executive: "In the seed business today we're not selling someone just a bag of corn. We're selling them a whole system of farming."[5] As this is being written, investors around the world are scrambling for a piece of the biotechnology pie. Meanwhile, desperate American farmers are lobbying Congress for relief from prices as low as many can remember.

It is here that Jefferson, I think, was ahead of Cochrane in his thinking. What are now called "intellectual property rights" were handled very simply by Jefferson. Knowledge could be given to one person without taking it away from someone else—the image of a person lighting a candle from that of another is sometimes used. Putting his ideas into practice, Jefferson never sought a patent for his advancements in plow design. While Cochrane worried for decades over technology, he was virtually blind to the issue of its ownership. Meanwhile, the corporations that brought technology to farmers used patent laws to dramatically revise Jefferson's ideal—you can light your candle from mine, but a substantial fee will be charged. Then, with biotechnology and extensions of patent laws to include basic life forms, the formula was modified again. Now, you cannot light your candle from mine at all. Instead, if you can afford it, you can rent a space in my lighted room. Otherwise, you must live in darkness rather than violate my ownership interest in fire.

Patent laws most clearly threaten farmer independence when applied to genetics. In the beginning, farmers selected seeds from crops they had grown and used those seeds to plant the next year's crop. Ownership, and the decision of which plants should be used for seed stock, was completely in the hands of the farmer. Then, with the rise of scientific breeding methods too complicated to be reproduced on most farms, ownership of the most desirable seed genetics passed into the hands of private seed and chemical companies, some of which have now attained multinational dimensions. These corporations, fortified with biotechnology and extreme patent protection, now place both farm profits and farmer freedom at grave risk. A modern farmer wanting biotech seeds will most likely enter into a contract with a seed company to use, but not own, the seeds.

The seed company will dictate production practices, marketing, and perhaps use of chemicals the seed company also sells. The farmer signs away rights to use the seed in later crops and pays a seed cost that, in some cases, includes an additional "technology fee" that further insures any profits will go to the corporation, not the farmer.

With many types of animal production, the story is the same. Genetics for types of farm animals are controlled as tightly as with seeds. Corporate control of genetics, in turn, translates to control of the farmers that use those genetics. So, why would a farmer enter into such an arrangement to begin with? Why not buy some less-controlled product somewhere else? The answer is that the corporate strategy to control farming goes beyond owning genetics: there will be no "somewhere else" offering viable alternatives. In some cases, providers of one genetic source assure sales by buying up competitors. For example, in Dain-Bosworth's 1997 analysis of the U.S. seed industry, one reads, "we continue to believe that only four or five seed corn companies will ultimately survive to serve the world market." In other cases, the genetics providers control product markets so tightly that farmers using other genetics will not be able to sell their product. Either way, the farmer's "just say no" option is effectively eliminated.

I have sometimes thought that the shortest possible economic history of U.S. agriculture during the twentieth century would be this: nonfarmers learning how to make money from farming. I suspect that the displaced steelworker, watching news of record stock market gains while trying to make ends meet with a minimum-wage job, might think this history could be applied more broadly. In any event, Cochrane's prediction that the profits from an industrialized agriculture would go to nonfarm investors, not to farmers, was right on the mark. This, in turn, led to the rise of economic superpowers in the food system. Jefferson so feared such developments that he proposed a constitutional amendment to prohibit monopolies; Cochrane was so concerned about the food system falling into the hands of big business that he would stop at nothing, including operating the entire food system as a public utility, to prevent it. In the minds of both men, the freedom of any country that does not have full and democratic control over its food supply is in great peril. During the course of this book I will therefore sketch the development of four corporate superpowers that rose to power as family farming declined—Cargill, Pioneer, John Deere, and Monsanto—and beg that the likes of Novartis, Continental, and Archer-Daniels-Midland will forgive me for not choosing them instead.

Cargill is the largest privately held company in the United States, three times bigger than the next company on the list. It has sixty-three thousand employees working in forty-seven businesses in fifty-four countries on six continents. The company's reach defies comprehension. The same company is the world's biggest grain trading company, the largest processor of cocoa in Europe, the largest pet-food processor in Argentina, and the first foreign company to set up an animal-feed plant in Japan. They make frozen orange juice concentrate in Brazil, handle the lion's share of U.S. grain exports, and supply parts to Chrysler and Boeing through their steel plants. Revenue, that is, gross sales, still comes mostly from trading anything and everything that can be called a "bulk commodity." But the profits, more and more, come from taking over the processing of these commodities into higher valued products. The total package amounts to $50 billion per year or more in sales. The total sales of all U.S. farms, big and small combined, are barely four times as large. Still growing, Cargill was granted federal approval in 1999 to buy archrival Continental's grain operations.

Pioneer Hi-Bred International, the world's largest seed company, was bought by chemical giant DuPont in 1999. Even without DuPont's enormous resources, Pioneer controls almost 40 percent of the U.S. seed corn market, has research facilities in thirty countries, and has annual sales of over one billion dollars. The company was founded with the discovery and development of hybrid seed that could not be reproduced and saved by farmers. Its growth was especially phenomenal during the "plant fence-row to fence-row" euphoria of the 1970s. Thanks to some remarkably high-yielding corn varieties with names such as "3780," Pioneer's sales grew five times as large and its profits eight times as large in less than a decade. Looking ahead, Pioneer sees great opportunities in the 40 percent of the world's corn plantings that do not yet use hybrid seed. And, thanks to their own efforts in biotechnology in combination with those of DuPont, their place at the technological table seems secure indeed. Today, a dwindling number of independent farmers watch television news stories of farm prices busted again by bumper crops, then commercials from Pioneer touting "technology that yields." For those farmers who can no longer afford to remain independent, the company's Optimum high-oil corn program stands ready and waiting with contracts to grow designer corn on company terms.

John Deere's company claimed the number one spot in farm equipment in 1963 and never looked back. During the 1970s sales increased fourfold and billions were spent in research, development, and capital improvements. Deere CEO William Hewitt's leadership did not go unrecognized—when *U.S. News and World Report* published their "Who Runs America?" issue in 1981, he was the only agribusiness executive listed. During the farm depression of the 1980s Deere stayed ahead of its competitors by developing its finance division, its lawn care business, and foreign sales. As a result, it came into the 1990s stronger than ever. The modern Deere and Company is an $11-billion-per-year operation featuring enormous tractors and combines with on-board computer systems. With thirty-four thousand employees and business in 160 countries, Deere dominates not only the U.S. market but that of the world, too. Farmers routinely finance their Deere tractors with Deere credit, making the company one of the twenty-five largest nonbank lenders in the United States. And, lest we think Deere is all work and no play, the Professional Golf Association named Deere as its official golf course equipment supplier in 1997.

In the 1970s Monsanto provided an interesting counterpoint to a popular movie of the time. In *The Graduate* a youthful Dustin Hoffman is advised by a wealthy friend of the family on how to make it big: "I have just one word for you—plastics." Monsanto was then big in plastics materials and synthetic carpet fibers, but already its agricultural products division was starting to steal the show. The drifting graduate might have been better advised with a different word: "herbicides." Agricultural chemicals, then biotechnology breakthroughs such as growth hormones for dairy cattle and herbicide resistant crops, became so profitable that Monsanto sold off its traditional fibers, plastics, and lubricants businesses to redefine itself completely as a "life sciences" company in 1997. The integration of seeds, chemicals, and biotechnology would lead their research and development program. At the same time, markets would be locked up by a program that spent billions buying out competitors. All of this was intended, in the words of one company executive, to "help transform the delivery of health, nutrition, and wellness to everyone on the planet." On a more ominous note, the interview also pointed out that, in Monsanto's view, "food is a very important currency."[6]

In the last few years even the largest agribusiness corporations have been merging into giants capable of unprecedented control of the food system. Phrases such as "dirt to dinner" and "plow to plate" are common among discussions of merger strategies. *Feedstuffs*, once the champion of anti-Cochrane communist baiting, now calls itself "The Weekly Newspaper for Agribusiness." In a single issue, that for May 18, 1998, the page one headlines caught even the most jaded of readers off guard. The lead story was Monsanto's $4.3 billion purchase of DeKalb and Delta & Pine Land, two of the country's last sizable independent seed companies. To the left of that, a joint venture between Cargill and Monsanto was heralded. The center of the page headlined Continental Grain's purchase of a controlling interest in Premium Standard Farms, thereby "laying the foundation for one of the largest integrated pork companies in the U.S." Just to that story's right, two of the largest cooperatives in the country were announced as entering joint venture discussions. As an ironic footnote, the bottom lead on the front page told of corn prices projected to reach four-year lows; more bad news for farmers. All this in one week.

Secretary Wallace ended his address at Monticello with a reminder that we could not go back to Jefferson. The world had changed too much for that. Instead, Wallace said we must go forward to Jefferson, searching always for the right balance in government, one gentle enough to respect individual liberties and strong enough to carry out the collective will of its people. Going forward to Jefferson in the new century will be more challenging than ever. Even the time-honored guideposts of market competition and strong rural values are lost among the sound and fury of big business scrambling to dominate a new world order.

A new agriculture and food system policy, whatever it might be, must work in a world where very large farms are sandwiched between much larger corporations that are powerful enough to control markets. Economists often lose confidence in the free market system when more than 40 percent of a market is controlled by its top four companies. Here is a partial list of agricultural industries that fail this test: fertilizer, herbicides, broiler processing, beef slaughter, pork slaughter, sheep slaughter, flour milling, soybean crushing, dry corn milling, and wet corn milling. Mergers and acquisitions also serve to concentrate private research and development into fewer and fewer hands. For example, DuPont and Monsanto had combined expenditures of $1.7 billion for research and development in 1996. For comparison, that was thirty-four times more than the Minnesota legislature gave its university's College of Food, Agricultural, and Environmental Sciences for all of its teaching, extension, and research programs combined.

Just as we can no longer depend on competition in the design of policy, rural values have also been swept aside. We have most often assumed a set of values among farmers that are only partly driven by profits. Farmer decisions are also influenced by resource conservation, by concern for rural communities, and by other objectives held desirable by citizens at large. At one extreme, farmers are the salt of the earth and will take care of our resources if only we will let them; at another, they are businesses like any other. Even then they are local businesses with local concerns. But when off-farm interests dominate farm decision making, the values of farmers are largely irrelevant. The values of off-farm investors drive decisions, and there is no guarantee that those values include the broader public interest. For example, cigarette companies are facing stiff legal challenges for disregarding public health. Those same corporations are among the nation's largest food processors. Philip Morris, home of the Marlboro Man, is better known to nonsmokers as Kraft, Maxwell House, and Oscar Meyer. RJR Nabisco brings us Joe Camel, Oreos, Grey Poupon, Cream of Wheat, and LifeSavers.

Where, then, shall we look for guidance? The life and times of Willard Cochrane have been those of a nation struggling to go forward to Jefferson amid changes our third president could hardly have imagined. That drama, played out on the stage of the American family farm, began in a golden age of loving grandparents, magnificent passenger trains, and a small farm tucked away in central Iowa.

2

The Golden Age

Willard Cochrane was seven years old when he first visited his grandparents' farm. During the summer of 1921, he and his mother began the long trip from southern California to Iowa in grand style on the Santa Fe Chief. He spent most of his time in the observation car, sometimes marveling at the deserts and mountains, sometimes watching wealthy businessmen play cards while he waited impatiently for calls to the magnificent dining car. They changed to a less-memorable train in La Junta, Colorado, and then in Omaha to one he can scarcely recall at all for the final run into Stuart, Iowa. Aunt Annie and Uncle Willie Wilson met them at the station, and soon they were all enjoying a huge breakfast featuring fried chicken and pies. Outside, a horse and buggy stood ready to provide travel on the last nine, dusty miles to a farm some forty miles west of Des Moines.

A farm typical enough of its time, it was about one hundred acres with a small river through the property providing water for the animals. What land his great-grandfather had cleared of trees and tall grass prairie, maybe sixty acres in all, grew corn for the pigs, oats for the horses, and hay for the cows. The remaining land was about half pasture and half forest. Grandfather Chambers occasionally proposed selling some of the valuable black walnut trees when money was scarce, but Grandmother would hear nothing of spoiling the place of so many Sunday picnics.

Ten or so cows provided milk for cream, which they sold, and skim milk for fifty or more pigs, the family's other big source of income. Grandmother had a huge garden and worked through the summer to can enough vegetables to fill a small cavelike area under the house. The beet crop was especially good that year, and there were always lots of potatoes. Memories of the famine that drove

The Golden Age

their own Scotch-Irish parents and grandparents into America were still fresh, even after two generations. Arriving too late to homestead their land, they bought enough from the railroad companies to get started.

Grandfather Chambers was a good farmer. His own father had never enjoyed farming; for all practical purposes he turned things over to his son when the boy was twelve. While the son farmed, the father grubbed stumps and did odd jobs around Adair County, an arrangement that seemed to serve everyone well. Grandfather Chambers held the farm together during the tough times of the late nineteenth century, then had the good fortune to farm in one of the best times American agriculture had ever seen. Before the war, prices had been strong and stable. During and shortly after the war, increased exports of food sent prices through the ceiling. Hard work and good farming paid off handsomely. This was, as Cochrane would later call it, "the Golden Age of American agriculture."

But then, as now, there was more to farming than growing crops and livestock. A successful farmer is most often a good, or lucky, speculator in the land market. Buying land at too high a price will sink even the best of farmers, and land with little or no mortgage will save most anyone. As grain prices went up, the cost of land in Iowa more than doubled between 1914 and 1920. Some, but not all, of Grandfather's neighbors saw that the Golden Age would be even better if they farmed more land. The rest were quite willing to sell if the price was right. As a result, two things happened. First, many sold their land to neighbors and moved West, lending truth to the saying that Long Beach was settled by Iowans. Then, the remaining farmers confidently took on new mortgages with every intention of paying them off during an eternal Golden Age. These larger farms were not nearly as self-sufficient as Grandfather's place since the additional land they financed was devoted to grain for export. This made sense as long as prices held up.

The Depression came early to the Midwest. Prices fell to prewar levels and below, leaving nothing to pay for new mortgages. When Cochrane came to visit again in 1926, his grandfather told him that all but two of the many farms between his place and town had been foreclosed. Grandfather had survived not by being such a good farmer but simply because he didn't buy more land during the boom.

The University of Minnesota kept records on many farms like the one Cochrane visited as a boy. The Minnesota studies show a striking contrast in labor requirements for farming then and now. For example, raising an acre of corn with horses and mules took over twenty-six hours of labor; by contrast, modern studies usually put that number at less than three hours. The early farms were also different in that most of the money gained from crop and livestock sales stayed on the farm. The farm family provided labor, and seed was saved from previous crops. Horses were tended by the farm family, fed from farm fields, and most likely bred from animals already on the farm. Fertilizer came in the form of manure from animals on the farm, and agricultural chemicals had not yet been introduced. Machinery was horse-drawn and relatively inexpensive—about $2.20 per acre according to the Minnesota studies. Hired help was partially compensated by room and board, rather than cash.

This was, for all practical purposes, the mythical American family farm. The family worked long and hard, was largely self-sufficient, and deserved a fair shake for feeding the rest of us so well. And if, for some unfortunate reason, the family fell on hard times, it was in everyone's best interest to help them out. The clear way to do this was to somehow see that what they were growing sold for a higher price. Virtually all of the resulting higher revenue would go right where it belonged, into overall pockets throughout the heartland.

The largest farming cost in the Minnesota studies was five dollars per acre for land. The mythical family farm, however, was one in which farmers owned the land they farmed. The land cost was, therefore, something for economists, not farmers, to worry about. White settlers had been lured West by the promise of cheap and plentiful land, but by 1900 all the best land was spoken for. This led to rising land prices, speculation, and more farmers renting land they would otherwise have owned.

A. H. Benton, in what was then the University of Minnesota's Division of Agronomy and Farm Management, published a bulletin in 1918 called "Farm Tenancy and Leases." It began with a broadside against absentee land ownership that would be unthinkable in a modern experiment station publication. Speaking of tenancy, he said: "At first this practice attracts no attention, but as it becomes more and more common the evils begin to appear. In all communities where tenancy has developed to any extent the effect seems to be bad. Soil fertility is being more rapidly depleted, farm and civic improvements are greatly retarded, and the social life of the community is not satisfactory."

In addition to Dr. Benton's concerns, absentee ownership of land created a way for nonfarmers to make money from farming. Then, as now, the high costs of renting and buying farmland often made owning farmland more lucrative than farming it. In the decades to come, technology, too, would transform

The Golden Age

farming from a business of labor into one of investment and high costs for seeds, equipment, and chemicals. When Cochrane was a boy, however, the corporations that would help bring about this change were mere specks on the horizon.

For example, the Golden Age of Agriculture was anything but golden for Cargill. The company's ventures into railroads and western irrigation projects had gone so badly that a creditors' committee was busily selling off holdings in land and timber. Bankruptcy, not world domination of agricultural interests, seemed the more likely outcome.

Pioneer Hi-Bred did not yet exist. In fact, the hybrid seed corn upon which the company was built was little more than a research concept when Cochrane was a boy. Back then, farmers selected seeds from corn they had grown and chose the best looking ears to show at the county fair. That farmers ever would buy seed bred by scientists using selection and inbreeding techniques farmers could not duplicate seemed unfathomable at the time. If someone could break down the farmers' resistance, a fortune was there for the taking: scientifically engineered seed could not be saved from previous crops and replanted year after year. Farmers would always have a seed bill to pay when they grew hybrids.

Meanwhile, the board of directors at John Deere was locked in a bitter struggle with their chief executive officer over whether they should produce tractors of any sort. Many thought farmers would spend money on automobiles long before they would give up their horses and mules. In that case, Henry Ford would be in an ideal position to make and sell tractors when, and if, they caught on with family farms. The situation was finally resolved in 1918 when Deere, the world's leading supplier of steel plows, purchased the Waterloo Gasoline Engine Company, manufacturers of the Waterloo Boy two cylinder tractor line.

Down in St. Louis, an enterprising Irishman named John Queeny was struggling to keep his new chemical company going. He had given the company his Spanish wife's maiden name, Monsanto, and hoped that saccharin production and sales would help him make good on the five thousand dollars he had sunk into the company. Soon there was caffeine, then vanillin, followed by a bonanza from World War I chemical sales that put Monsanto on the corporate map. While Monsanto had no involvement in agriculture during those times, one of the company's chemists unknowingly foretold the future of family farming: "We will succeed only to the extent we can provide for each customer the kind of technical assistance he needs in order to make our products do his job."[1] Soon enough, products from assorted nonfarm companies would be doing the farmer's job in ways no one, least of all the farmer, could have imagined.

The Golden Age

A chain of events was put in place in 1908 that, in its own roundabout way, resulted in Willard Cochrane being born in California rather than Iowa. One of the Chambers boys was never very healthy. In the fall of 1908 the family doctor said Clyde would not likely live through another prairie winter. The doctor had recently visited the Imperial Valley of California, and now he offered it as the boy's best chance of survival. Grandmother and Grandfather Chambers soon announced their intention to spend the winter out West with Clyde. Clara, one of their daughters, insisted on going, too. At long last, after Clara promised to sell her only cow to pay her way, her parents gave in.

Such determination must be hereditary, for the son she later gave birth to became widely regarded as having an all-but-terminal case of it. Her father, too, appears to have been a carrier of the determination gene. He was a bit of a sleepwalker and was far too frugal to consider a sleeping berth on the train west. One night he rose from his uncomfortable chair in the passenger car and, deeply asleep, walked out the door of the train as it sped through an Arizona night. He awoke rather abruptly, badly bruised but able to walk, and set off on foot to the nearest station, where he boarded the next train to rejoin his anxious family.

Clara was taken with California life and stayed on after the others returned to Iowa, working as a telephone operator in El Centro. Meanwhile, Willard W. Cochrane I was running a big cattle operation near the Mexican border. He sometimes came into El Centro to make long-distance calls to his bosses in Los Angeles, and so their paths crossed. They were married in 1913 and the following year had a son, Willard W. Cochrane II. A few years later, the senior Cochrane and his bride started a farm typical of California, unlike anything Clara had seen in Iowa. It was a dairy, but a large one even by some present-day standards. Just north of Van Nuys, the Cochranes and a crew of hired Portuguese immigrants milked 120 cows.

It was always a golden age of one sort or another in California. There were gold rushes, land rushes, great conditions to grow and market farm products, and a climate that was beginning to lure new residents by the thousands. Even in bad times, everyone knew that the constant flow of newcomers would sooner or later bring renewed prosperity. But when Clara and her son returned from their visit to Iowa in the fall of 1921, their faith in the California good life was severely shaken. During the summer their house in Van Nuys had burned to the ground, leaving family pictures in ashes, fine silver settings melted into shapeless lumps, and Mother's beautiful piano in ruins. Will's father, in their absence, had decided that a tent house would be adequate for their needs, and that is where they lived for the next year. Cochrane was too young at the time to remember if it were due to financial reasons or family stress, but the dairy never recovered. In 1923 his father decided to return to the world of managing the large corporate farms that have always dominated California agriculture. He signed on with Fontana Farms, a huge operation fifty-five miles east of downtown Los Angeles.

Fontana Farms owned thousands of acres. Its primary business was luring adventurous souls from around the country with the promise of living the good life in exchange for tending a few thousand laying hens. Fontana Farms stood ready to provide land, birds, and marketing services for eggs. Fontana Farms was also big in vineyards and orange groves. Growing grapes and oranges on the sandy soils of southern California required a huge supply of fertilizer, for which animal manure was the only practical option. As a result, Fontana set up a hog ranch that produced both meat for sale and a constant supply of fertilizer to fuel the lucrative vineyards and orange groves. The hog ranch would make the most ambitious of modern farmers proud. Some eighty thousand hogs dined in high style on ten or twelve railroad carloads of Los Angeles garbage each day. Their manure production was equally voluminous, but careful attention to prevailing winds in its siting kept the ranch from being a problem for the land sales division.

Cochrane's father, perhaps because of his background with livestock, was hired to manage the hog operation. He, no doubt, had many of the same reservations any cowman would have had about being thrust into the company of hogs. For young Cochrane and his mother, however, Fontana was all good news. Cochrane remembers the Fontana Farms model community as a beautiful place, covered with countless new orange groves and miles of eucalyptus windbreaks. There was no smog to block the view of the snow-covered San Gabriel Mountains, and the occasional trip to downtown Los Angeles took less than two hours on the Big Red electric trolley cars. His mother got a new company house and took a job in the land sales office. Meanwhile, her young son was free to pursue his "Riders of the Purple Sage" imagination on a nearly full-time basis. There were plenty of open spaces to roam and play, orange groves to camp out in, and even a horse his father had kept along with a few cows when he moved to Fontana. For two glorious years, Cochrane rode his father's horse among the hills south of the hog ranch and read cowboy books from the new town library.

Then, in 1925, Cochrane's father decided it was time to move on. The good salary and nice house gave way to a huge debt for rebuilding the dairy herd. For a while, it looked like the new venture not far from the hog ranch might actually succeed. Young Cochrane was eleven now and, to his great pleasure, was able to work with his father on the dairy. He fed calves, washed cows to be milked, and best of all, got to help with the milk route in the Fontana Farms area. In exchange for running milk from the truck to customers' houses, he was allowed to drive the Model T on the back roads and around the farm.

The dairy also gave Cochrane and his father some time, at long last, to get to know each other. While many knew the elder Cochrane as a hard, difficult man, for Cochrane he was a great companion on trips to San Dimas to buy orange pulp to feed the cows, to Victorville to buy hay, and to a ranch east of San Bernardino where the dry cows were pastured. All this changed on March 21, 1926. Cochrane was sitting at the dining table doing homework after dinner. His father was sitting nearby in an easy chair. Without warning, the father cried out in pain from a massive heart attack. While a terrified son tried to comfort him as best he could, Cochrane's mother set aside her Christian Science principles in frantic search for a doctor. Hours passed before one could be found, and by the time he arrived there was little to be done except to take care of paperwork for the death certificate.

Thus began the Great Depression for Will and his mother. A farm sale followed his father's death and virtually everything that had constituted the boy's life was sold—cows, land, equipment, horses, a wonderful Studebaker touring car with a California top, and the Model T truck he had enjoyed driving so much. Even so, there was not enough to cover the farm's debt. His mother had to dip into the life insurance money; the remainder she tucked into the local savings and loan. When the bank went under, Cochrane and his mother were all but penniless.

This was the last time the future presidential advisor on agricultural policy lived on a working farm. Because of Iowa, Fontana, and riding with his father on the dairy route, Cochrane would always see farms as places of great happiness and security. He would also know, forever, how financially fragile a farm could be, and understood firsthand the personal tragedy that occurred when a farm came undone.

Country life quickly gave way to city. Cochrane, now twelve, and his mother found a small apartment on Vermont Avenue in Los Angeles in the fall of 1926. Shopping was nearby with easy access to streetcars, two items of great importance to a family unable to afford a car. In fact, with thirty-five dollars per month for rent taking half of his mother's pay, there was little of anything they could afford. Alone, because he knew absolutely no one in the city, and fearing that his mother, too, might die and leave him completely helpless, a miserable young man watched the days slip by from the window of the tiny apartment.

His mother worked during the day and spent evenings either helping her son with his studies or trying to teach him the ways of Christian Science. Slowly, Cochrane discovered the nearby Ravena Theater with its cheap matinees and the local library with its stock of adventure books. Mr. Moore, the apartment building manager and owner of the unpainted furniture store across the street, occasionally found small jobs for his new tenant. And, to Cochrane's pleasant surprise, the nearby Virgil Junior High was not as bad as he had imagined it would be. Best of all, his joining Boy Scout Troop 113 brought a way to meet friends, camping trips into the nearby mountains, and Scoutmaster J. C. Earle, who came to stand in for the father Cochrane so often missed. Within a year of moving to Los Angeles, Cochrane had his Boy Scout group and a best friend in Ray Edwards and was no stranger at the local playground. There was even a *Shopping News* route, which put three dollars into his pocket every week.

Junior high, then Los Angeles High School, passed by with no indication of the scholar Cochrane would one day become. Although a standout in history, geography, and English literature, he struggled with physics and foreign languages. Then, as now, he was lost without a dictionary to help with spelling. He put in a good effort in both sports and attracting girls but, alas, showed promise only on the former front. Stories of his high school days are relatively few and limited to an occasional strapping by the vice principal for boys and the fortuitous discovery of a small crack in the wall of the girls' locker room. His memories of being a Boy Scout, however, are crystal clear. The late-night climbs up Mt. Wilson and views of the lights stretching across the Los Angeles basin to the sea, skinny dipping in deep secluded pools on hot summer days, camping and backpacking among the rattlesnakes, and becoming an Eagle Scout are the stuff of which those days were made.

When Cochrane graduated from high school in June of 1931, he decided to work for a year. He had hopes of improving his running ability enough to be on a college track team and maybe finding some inspiration on a course of study. Instead, he got a firsthand look at life in the Depression, something that would affect him deeply for the rest of his life. He spent his first month after high school looking for work. His mother, meanwhile, continued to drag him to Sunday school with waning hopes that her efforts would someday bear fruit. As fate would have it, the Sunday school teacher had a small factory in East Los Angeles and offered Clara's son a job making card tables and other cheap furniture.

The Sunday school teacher was a very different man the other six days of the week. He ran a sweatshop where a stern foreman pushed a crew of fifteen teenage boys in what Cochrane regarded as a decidedly unchristian manner. Pay was by piece rate, so many cents for each part assembled. No matter how hard the boys worked, their weekly pay never topped twelve dollars. The more they assembled, the lower the piece rate became. Cochrane lasted two months, then quit in disgust. Almost seventy years later, Cochrane still complained bitterly about this job. It was, in some ways, a foreshadowing of the "treadmill" he would later propose as a way to think about the family farm's economic dilemma. That a Sunday school teacher, of all people, should run such a shop was something for which Cochrane would never forgive organized religion.

Another month of wandering in search of a job passed until a Boy Scout connection led to something much better. Typographic Service Company hired Cochrane as a bicycle messenger for fifteen dollars per week. Officially, he delivered ad proofs around downtown; unofficially, he made sure no typesetter was unduly forced into sobriety by Prohibition. He felt fortunate to have avoided serious injury in the traffic and, even more so, to have landed a job in a union shop. People working together could, even under difficult circumstances, do much better than they could by competing individually on the Sunday school teacher's treadmill.

With all the bicycle riding and evening runs at a local high school track, things started to look up. He got his time for the mile run below five minutes and could do the 440 in a respectable fifty-eight seconds. By the fall of 1932, he was ready to start college.

Cochrane may have been ready to go to college, but the University of California system was not yet ready to take him. There was the matter of his grades and course work to consider. Accordingly, he began his studies at nearby Los Angeles Junior College in search of an A or B in a year course in science and in a year course in a foreign language. All of this was intended to someday lead to a successful career, first as a track star, then as a coach. In his first collegiate cross-country season, Cochrane ran number three on a ten-man team. Then in the spring he tried out for his favorite running event, the 880, but was not fast enough. Instead, he lettered in the mile and worked to become a straight A student in his freshman year.

He tried the drinking and partying life of late Prohibition and, after giving it a fair chance, chose instead a steady relationship with a young woman of strict Methodist upbringing. She put a quick stop to the drinking and favored dancing, picnics, and trips to the mountains. Cochrane has often talked with me of those days and hardly mentions the Depression with its constant trail of low-paying jobs from janitor to pot scrubber to casket maker. For him, Los Angeles was, above all, a wonderful place to be young and in love. In his second year of junior college Cochrane ran on the winning two-mile relay team at the Fresno Relays. Brutus Hamilton, the coach of the California track team, was watching and invited the would-be star to come to Berkeley.

He began life at Berkeley as a fraternity man, a role he never felt comfortable with. He disagreed with Coach Hamilton on his natural track ability and slowly lost interest. Then his fiancée began studies in chiropody at a college in San Francisco, which set in motion a long road to a broken engagement and the parting of their ways. In the midst of all these setbacks, Cochrane did as so many before, and so many more to come, have done: he picked up a college catalog and hoped for inspiration. A career in track and coaching was out of reach, and his intellectual love, history, offered no way out of his financial straits. He thought of his farm background, and that he had done well in introductory economics, so decided to pay a visit to E. C. Voorhies, the undergraduate advisor for agricultural economics. He was immediately taken by the gruff professor who told him how things were changing in Washington.

The United States Department of Agriculture had long been devoted to providing research and education on improved farming techniques. Agricultural economics, and what we now call farm policy, were all but unheard of in the department's earliest years. This situation had changed dramatically, however, by the time Cochrane leafed through his catalog in 1935. Henry C. Wallace, one of the Corn Belt's most famous sons, was secretary of agriculture. Wallace was far more inclined to think that government oversight of production had its place. One of his first acts was to establish the Bureau of Agricultural Economics in 1921, just in time for an incredible convergence of events, which forced the government's hand on agricultural policy.

In 1920 one acre in four of harvested crops went to feed the horses and mules that helped produce it. Within a decade, the conversion to tractors would be well on the way and most of that land would be producing more grain than the economy knew what to do with. On top of that, U.S. farmers had been advised during World War I to "Plow to the Fence for National Defense" and had brought 40 million new acres under the plow in the "Food Will Win the War" campaign. There was no obvious way to slow down production at the end of World War I, and a massive farm depression was soon to result. Finally, the 1920 census brought news that any efforts to solve farm problems by the government would happen in a country that for the first time in its history had a larger urban than rural population.

The Bureau of Agricultural Economics was the U.S. government's largest economic research and service organization. Professor Voorhies thought the eager young student before him might well find work with this group or one of the many other agencies and programs serving agriculture at the time. That he would someday lead the programs was, I suspect, more of a stretch. Nonetheless, Cochrane and agricultural economics were a perfect match. He sailed through courses in agricultural economics, agronomy, soils, and horticulture and was so engrossed in his schoolwork that a potentially miserable time was made tolerable. In 1937 he graduated second in a class of eight agricultural economists.

He had been interested in graduate school for some time, but when his application for an assistantship at Iowa State was rejected his thoughts turned toward working in the California orange business. Then, a week before graduation, Professor Voorhies called to say that he could get him a position with a new program at Montana State if he was interested. Cochrane quickly got his mother's approval, looked up Bozeman on a map, saw that it was in the mountains he loved so well, and accepted the position on the terms that Voorhies had worked out on the phone that very day.

A story about an unfortunate sheep farmer circulated during the time Cochrane was in graduate school at Montana State. The western rancher sent a load of sheep to market out East and impatiently waited for the check to come. Instead, he received a letter explaining that the sheep had not brought enough at market to pay for the freight to get them there. An invoice for the difference was enclosed. The farmer was a proud man, too proud to admit he had no cash to pay what he owed. To make things right, he sent another load of sheep.

This story, and many more like it, was born of the "farm problem" that would dog Cochrane and everyone else in farm policy for the rest of the century. American farmers routinely produced more food than American consumers were willing to buy at reasonable prices. Surplus production and low farm income had already led to government farm programs that were not terribly different from those Cochrane eventually challenged twenty years later. To reduce production, the government supported farm prices in exchange for farmers setting aside some of their land. The idea was that lower production would lead to higher farm prices, so the programs would never be very expensive. Dream on.

Cochrane, as much as any agricultural economist, is associated with the liberal notion that government has a legitimate and helpful role to play in the farm economy. The idea that the government was there to help people was reinforced at Bozeman in two important ways. For one, the agricultural economics department was in a building that had an entire floor devoted to a WPA project. Cochrane was never exactly sure what the people who worked there were doing, but he knew that the public jobs they were given made all the difference in the world to them and their families. Most would otherwise have moved on in desperate search of employment, and some might not have survived the depths of the Depression. Cochrane also gained firsthand knowledge of the government helping those in need through his research project. He studied farms throughout Montana to find out what made some more successful than others. The income statements of every successful farm showed government programs helping crop and livestock sales. Cochrane also noted that the successful farmers made liberal use of loans from the Federal Land Bank. In fact, "willingness to cooperate with governmental agencies" was one of the main characteristics of successful farmers noted in his thesis.

The farms Cochrane studied were just similar enough to his grandfather's farm to be familiar and just different enough to show that the stage for enormous change was being set. The farms were familiar in that the farmers grew most of their food and owned the land they farmed. Most of the labor was from the family, and horses still helped out with the chores. But things were changing. The farms were more specialized in what they grew for sale; machinery was making much larger farms possible. The successful crop farms he studied averaged 450 acres in wheat alone. The small cost for horse drawn machinery had become a big cost for tractors: fuel, repairs, and borrowed money to buy them. Seed dealers wanted their share in exchange for new high-yielding varieties. The farms were taking more money in, but they were paying more out, too. In addition, the new ways of farming were hardly good for prices. More tractors meant fewer horses and mules, so land that once grew their feed would now dump food onto the market; new seeds meant higher yields.

When Cochrane summed up the characteristics of the successful farmers he analyzed for his master's degree, the first thing on the list was that they were hard workers. They no doubt were, but no data were collected to give any idea whatsoever of whether successful farmers worked any harder than unsuccessful farmers. The conclusion came not from his research but from the author's own value system, one apparently shared by his faculty advisors, for the conclusion went unchallenged.

The Golden Age

That hard work leads to success was but one of several values Cochrane, like others of his generation, brought into the farm policy debate. Another was that the family farm was a good thing, a place of security in an otherwise treacherous economy filled with sweatshops like those the Sunday school teacher ran. If people would work together, as with the union in the bicycle delivery job, they could better themselves. The government, too, could play a strong role in helping people help themselves. When it came to farming, the government could help control surplus production and guarantee good prices for farm products. Progress, above all else, was essential—his successful farm study concluded that the good farmers were "sufficiently progressive to accept new and promising ideas and practices."

These values, flowing so perfectly from the life of the young Willard Cochrane, are essentially those that shaped farm policy for the remainder of the century. Good in many ways, they are also contradictory. For example, working hard did not, and in Cochrane's judgment could not, improve his situation at the Sunday school teacher's sweatshop. Higher farm prices could not put money in farmers' pockets if the pockets had holes in them. And how could you patch the holes if they were brought about by progress? As for unions, they were for wage earners; farmers saw themselves as independent businessmen. Last, and hardly least, was the very idea of "surplus" food in a hungry world.

As we shall see, the careers of many people, including Cochrane, and the tax dollars of many more, have been spent trying to save something called the "family farm" that was hiding in these values and suffering from these contradictions.

3

The Treadmill

Cochrane graduated with a master's degree from Montana State University in June of 1938. He was twenty-four years old and starting to look like he might make it as an academic after all. He spent the summer grading oranges back in California, but his thoughts were in Minnesota, where he had been offered an assistantship in the Ph.D. program. In September he and his mother drove to Minnesota and stopped to visit the family farm in Iowa along the way. Grandfather Chambers had died several years earlier. Cochrane's favorite uncle, Zene Chambers, was running the farm. His farming left something to be desired, to be sure, but that was more than compensated for by his willingness to include his admiring nephew on expeditions into town in search of drinks and stories. Apart from its new management, the farm had not changed much. The same land was farmed in virtually the same way, with only a small tractor to signal what was to come.

Zene was a tragic man in some ways, a victim of the family farm tradition. That he was ill suited to farming meant little in a world where passing a farm on from generation to generation had the status of divine revelation. First, his teenage son committed suicide, then drinking became more and more of a problem, next gambling, and finally his wife left him. Grandmother Chambers forgave his rent payments when she could but finally gave up. She eventually sold the place to a neighboring farmer who wanted to get big enough to use some of the new farm equipment coming on the scene. As for Uncle Zene, he did all right once the farm was gone. He remarried, raised a second family, and went to work as a government farm program agent.

The land once owned by the Chambers family went on producing just as much, even more, once Zene was gone and a better manager started calling the shots. Such was the stuff of a great contradiction soon to come in farm policy. Within a decade there would be calls for "painful, but necessary" steps to get farmers off the land as a way to control rising surpluses. Predictably, it was inefficient farmers, not those who were efficient, who would leave. In order to control production by controlling the number of farmers, the efficient farmer somehow would have to be eliminated. No self-respecting free enterpriser would consider that.

Uncle Zene, bless his heart, wasn't the problem. He, and millions more like him, had stood guard for decades over a farm economy remarkable in its ability to preserve farm numbers. These farms sometimes made enough for the farm family to live well and sometimes didn't. But they never made enough to support the legions of nonfarm interests hawking improved farming methods. Something would have to give. What "gave" was the number of family farms. We often think of the twentieth century as one in which farm numbers steadily declined. Actually, farm numbers were stable in the first third of the century. The high-water mark of farm numbers in this country—6.8 million—came in the 1935 census. Every census after that recorded a smaller number of farms, with fewer than 2 million remaining in 1997.

At the University of Minnesota Cochrane found more sophisticated courses in microeconomics and advanced statistics and a department influential enough to get him a job in Washington for work that could well lead to a dissertation. Two offers came within two days in 1939, one from the Farm Management group at the Bureau of Agricultural Economics (BAE) and the other from the Cooperative Research Service of the Farm Credit Administration. He had not grown up on a farm, so he thought he would never fit in well with the farm management group. Working with cooperatives seemed a better match for his background and political views. He took the Cooperative Research Service job in Washington with high hopes. What he would quickly come to view as the deadest group imaginable, one that would discourage him so much he became permanently disillusioned with the idea of farmer cooperatives being worth much, was waiting to greet him.

The young idealist, who came for Ph.D.-level discussions on the nature of farmer cooperation, soon found himself doing a statistical inventory of defunct cooperatives. In short, he was a clerk, a clerk under the supervision of an elderly man who at times appeared dead and at other times merely asleep. Bored with his work, Cochrane in desperation agreed to spend a summer in southwest Oklahoma teaching business principles and bookkeeping to managers of struggling cooperatives. That, along with the prospect of going back to his stultifying job in Washington, convinced him it was time to look for something else. The "something else" was first suggested by a couple of Harvard graduates at the United States Department of Agriculture (USDA) who told him of how well John D. Black took care of "his boys" at Harvard. Black was among the country's most well-known agricultural economists, and Harvard was, well, Harvard. Cochrane went to his first professional meetings at the end of 1939 to meet his future mentor. The two men liked each other from the start. With very little remorse, Cochrane decided that going to Harvard would be far better than returning to Minnesota in search of a way to make gold from the lead of his cooperatives research.

At Harvard, he laid the groundwork for his greatest intellectual accomplishments. Harvard, above all, was a place where there was no need to question the wisdom of government action in the economy. Disciples of Keynes were everywhere, and government action was simply assumed. The interesting question was which actions the government should take. Government management of the economy was born of the Depression and soon would be tested again during World War II. The new macroeconomics held out the prospect for a comprehensive view of an entire economic system. Cochrane was particularly taken with this idea and began to see that even the great John D. Black viewed agricultural policy as a series of unrelated solutions to problems of the day. There was no macroeconomic view of the agricultural economy to guide and unify farm policies. The seeds of Cochrane's 1958 masterwork, *Farm Prices: Myth and Reality*, were being sown.

At Harvard he first read Schumpeter's *Theory of Economic Development*. Here, at last, was a theory that appealed to the historian in him. Conventional thinking of the time regarded economic action as occurring in a static, timeless world. Actions taken today would have the same effect as those taken tomorrow. In Schumpeter's world the economy was a circular flow, which, from time to time, was raised to new levels by innovative entrepreneurs. They made profits, but those who followed earned less and less until there were no further profits to be made. This is the stuff from which the treadmill, Cochrane's most enduring idea, would be fashioned.

Ironically, Harvard not only educated one of our greatest agricultural economists but almost talked him out of the profession altogether. While Harvard may have been a premier school for general economics, it offered little beyond Black in the way of agricultural economics. The siren song of becoming a general economist played loudly for Cochrane in those years. History was always his first love, and Harvard's economic history program was a constant joy. He did well, too, in his economic theory courses; so well, in fact, that Black had him tutoring some of his other budding agricultural economists. His Ph.D. research dealt with the ways higher levels of food consumption might stimulate the economy as a whole. It barely mentioned farmers. Later, at the Bureau of Economics, he wrote a memo to staff on "Functional Finance" that was of such quality he was courted by the Treasury Department. Later still, at Penn State, he was hired as a professor in agricultural economics with the understanding that all of his teaching would be in general theory courses for the economics department. It wasn't until 1951 that he resolved his conflict by accepting a position in agricultural economics at the University of Minnesota and going home to the problems of the chronically ill farm economy.

Cochrane remembers his days at Harvard as a time of uninterrupted work and study. For the first time in a long time, he had virtually no social life. Even his car was sacrificed to pay tuition. He studied, he worried about the draft, and he studied some more. He worked without rest, summer sessions and all, in an all-out effort to finish his courses at Harvard before the military caught up with him. He gained one deferment, kept studying, and was within sight of his preliminary exams as he prepared for an economic history test one Sunday morning in December of 1941. The radio brought music, then the stunning news that Pearl Harbor had been bombed. There would be no more escaping the draft. Professor Black agreed to schedule his preliminary exams in short order and in January 1942 Cochrane emerged as admitted to candidacy at Harvard.

The day after he passed his exams at Harvard, Cochrane was on the train bound for Washington DC. He was hoping to move in with his old roommate, Charlie Coffey, who was staying at Miss Billie's boarding house on Sixteenth Street just above K Street. Charlie was already rooming with someone and there was no way to fit another bed in the small room. Miss Billie did the best she could and made a bed for the Harvard man in a closet that opened onto the hallway. He called that closet home for several months while he awaited a commission in the navy and worked for the Office of Price Administration on rent controls.

Cochrane immediately set himself to the task of resurrecting his social life. Clay Henderson, Charlie Coffey's roommate, offered to help out by setting up a blind date with a young woman he had known at the University of Arkansas. Mary Herget, after teaching school in Arkansas, had like so many others gravitated toward Washington during the war. Cochrane and Mary spent their first date at a fancy hotel in Washington, the name of which escapes him. He does, however, recall the champagne cocktails and that Mary was both very attractive and a wonderful dancer. One date led to another and in no time they were going steady.

In the spring of that year he was inducted into the navy. Instead of being sent to some training school for new officers, he was ordered directly to the naval code room on Constitution Avenue. Any glamour that might have been associated with being a naval officer eluded Ensign Cochrane in as thorough and complete a way as can be imagined. During World War II it was required that all coded messages be handled only by officers. There was none of the spy-story cracking of enemy codes, at least none he ever saw. Instead, he shared long tables with one hundred or more other junior officers in a large room. Each man typed endless, tedious messages into typewriter-like machines with special wheels for coding and decoding. The messages ranged from maddeningly complex latitudes and longitudes of enemy submarine sightings to monumentally boring requests for supplies.

The endless routine of twelve hours on, twenty-four hours off, in the code room sent many packing with nervous breakdowns to show for their efforts. Cochrane first felt the strain with a bout of appendicitis that landed him in the Naval Hospital in Bethesda. While he was recovering, another doctor recommended more surgery to remove some badly impacted wisdom teeth. Now thoroughly miserable, he spent a few more weeks in the hospital looking forward to Mary's daily visits. Cochrane proposed marriage on one of those visits. Mary, widely regarded as the enduring support for Cochrane's career, raised their four boys and edited virtually everything he wrote. I remember her planning and presiding over a river cruise in honor of his eighty-fifth birthday or plotting a humorous reception for the new neighbors. The years have done nothing to diminish the warmth of her smile or the laughter in her eyes.

Once out of the hospital, Cochrane was soon back in the code room, this time feeling the stress in progressively worsening pain in his back and legs. Finally the day arrived when he was so incapacitated that he couldn't get out of bed at all. An ambulance took him back to Bethesda. He made the best of his time in the hospital and soon discovered that a prominent member of Congress was languishing in a room down the hall. A few quick visits revealed that well-wishers and lobbyists dropped by enormous quantities of whiskey for the congressman. Rather than see the bottles clutter up the room, Cochrane led a group that spent every afternoon with the ailing politician and, in their own way, contradicted his dissertation research. High-level consumption might not eliminate food surpluses, but it worked wonders on surplus whiskey.

Upon his release from Bethesda, he was given a medical discharge from the navy and advised to seek a drier, more favorable climate in hopes that his arthritic symptoms would abate. As a result, he was assigned to the War Food Administration Office in Denver, where he had an experience few living agricultural economists can claim—he participated in the public management of the entire agricultural economy. Everything from retail prices to wholesale prices to farm prices to transportation was controlled directly from government offices. The task was unprecedented. As always, the domestic food demand had to be considered. But during the war there were also huge amounts of food going to the military and to the lend-lease program. Looking back, he still thinks the farm economy worked at its best during those times. Meat shortages and the like aside, food was distributed equitably and in a way that would lead to healthy diets. His comparisons of agriculture to a public utility were still far in the future, but it is easy to imagine them starting to take shape in the Denver office.

Cochrane wrote a long memo on the meat distribution problems in the West that became widely circulated. Word was out that a very bright, very draft-proof agricultural economist was looking for a bigger pasture. He was soon back in Washington working for the Bureau of Agricultural Economics. He was feeling better, too. A smart chiropractor discovered what Bethesda had missed: one of his legs was slightly shorter than the other. A thin lift in one of his shoes provided permanent relief from the pain that had cut short his navy service.

In describing the mid-1940s at the Bureau of Agricultural Economics, Cochrane wrote in his personal papers: "My association with H. R. Tolley, Bushrod Allin, James Maddox, James Cavin, Nathan Koffsky, John Brewster and others in this period make this the most exciting and rewarding work experience of my life and turned me into the unreconstructed liberal that I remain today." This statement seems all the more striking in light of the professorships, high honors, and association with President Kennedy that were to follow. As is often the case, though, one's greatest achievements often happen maddeningly early in life. Barely thirty years old, he wrote his most enduring journal article, "Farm Price Gyrations: An Aggregative Hypothesis." It received the profession's top award then and was once again honored during a tribute to his career some fifty years later. On top of this, he finished his dissertation and graduated from Harvard during the June commencement of 1945. As already mentioned, his Keynesian memo on "Functional Finance" brought suitors from the Treasury Department. Even the attention of Congress was gained by his work on the publication series "What Peace Can Mean to the American Farmer."

If there is a downside to such success, it is that the ideas that come with it sometimes stay longer than they should. Cochrane's dissertation research and other work at the BAE left little doubt that production controls would always be needed for U.S. agriculture. He noted many ways, such as school lunch and food stamp programs, that could be used to increase food consumption among low-income groups. He also thought export markets should be expanded whenever possible. But he never thought any of this would stay ahead of technology on the farm. A free market approach to farm policy would therefore never work, because there would always be more food than could be sold at reasonable prices for farmers. As we shall see, this idea became increasingly difficult to sell.

The BAE days also produced the first extensive file of his unpublished professional writings. The picture of a man at the top of his game, full of liberal euphoria from helping to manage the farm economy during the war years, comes through loud and clear. Any suspicions I might have harbored that his cantankerous personality was born of advancing years were convincingly put to rest. Consider his memo to staff announcing Harvard professor Alvin Hansen's upcoming seminar at the bureau that, in part, read: "This discussion should provide certain members of the Bureau who find themselves in disagreement with some of Mr. Hansen's ideas with the opportunity to 'refute' those ideas. In the interest of democratic procedure, however, not more than two persons will be allowed to talk at once, and 'Tommy' Thomsen will not be permitted to hold the floor more than one-third of the time." An outraged F. L. Thomsen rose to the bait and lashed out with a two-page diatribe pointing out that "any talk-a-meter recording events in our various conferences would find you gentlemen at least running me a neck-in-neck race in that regard." Thomsen hoped that the bureau would someday consider "measures in addition to the second-hand theories of intellectual parrots trained by an unrepresentative clique of maverick economists and sociologists."

Cochrane was at it again a year later with an unsolicited program for redirecting virtually all of the BAE's research program. Ever the politician, he ended with this advice: "The work should not be undertaken by a Bureau-wide Committee. Research is weakened by committee action when unanimous agreement is required. Innovations in research are always stifled by the less imaginative members of the committee—and the result is mediocracy." Still another memo from the same time took a subcommittee report to task on the grounds it was "too evenly balanced in argumentation to be of real value to policy makers."

His suggested research agenda, by the way, had "the problem of the impact of technological innovations in agriculture from the farm unit to the retail outlet" as its very first item. That would have to wait, however; the salad days of Cochrane and his liberal mafia at the BAE were numbered. The war was over, and it was time to get back to business and beat the vast swords of war into industrial-strength plowshares.

Deere and Company had this to say in their 1950 Annual Report: "It has been a John Deere practice to utilize advertising space in various publications—particularly trade magazines—for conveying to the public the humanistic spirit of the John Deere organization. Usually personal in tone, these messages have, for the most part, spoken in support of such venerated causes as Americanism, Free Enterprise, Soil Conservation, and similar subjects." The venerated causes of Americanism and free enterprise made great growing conditions for off-farm corporations.

Cargill recovered nicely from its early financial troubles. Throughout the 1930s the company invested heavily in facilities to store and transport grain. Then, when World War II made international grain trade very difficult, Cargill began to diversify and slowly take on its more modern form. In 1945 the company purchased Nutrena and entered the business of processing corn and soybeans. The resulting vegetable oil and high-protein meal led the company in two important directions, food for direct human consumption and feed for animals. By 1960 Cargill's annual sales had reached the billion-dollar mark.

Pioneer Hi-Bred had within a single generation sold nearly every corn farmer in the country on the idea of buying expensive hybrid seed rather than saving it from past crops. Why? So they could grow more corn than markets could ever absorb and thereby wreck prices. Credit for this marketing coup goes to Iowa's Wallace family. Henry C. Wallace was appointed secretary of agriculture just five years before Pioneer was formed. He also owned *Wallace's Farmer*, a progressive and widely read farming newspaper that stood ready to praise and advertise hybrid corn. His brother had been an agricultural advisor to President Theodore Roosevelt. His son, Henry A. Wallace, was already editor of the newspaper when he founded Pioneer Hi-Bred Corn Company in 1926. During the company's years of early growth, he also was secretary of agriculture, then vice president of the United States under Franklin D. Roosevelt. Year in and year out, Pioneer dedicated itself to an aggressive research program aimed at ever-higher-yielding corn varieties. The new seed was sold through a network of farmers selling to other farmers, an idea so successful that it was soon copied by every competitor. Pioneer and rival DeKalb were neck-in-neck for the number one spot in seed sales during the 1950s and 1960s.

Monsanto was a much bigger, much smarter company in the 1950s. They had learned how to make money, that "promotion has one objective: increased sales," and that research and development aimed at patentable products was the key to profits.[1] During the mid-1950s U.S. companies like Monsanto were producing 20 million tons of fertilizer, 40 million pounds of arsenic-based pesticides, 130 million pounds of DDT, and 30 million pounds of old-school herbicides like 2,4-D and 2,4,5-T each year. Monsanto, however, viewed agricultural products as a sideline to its main businesses of plastics and industrial chemicals.

The modern Monsanto began to take shape one morning in 1958 when the company's new president went down for breakfast at the Hotel Pierre in New York. He saw Frederick Hatch, recently retired from Shell Chemical's agricultural division, sitting alone and asked if he could join him. Thus began the plan to form Monsanto's Agricultural Products Division. Now everything was in place to combine research and development for proprietary chemicals, the capacity to manufacture them, and the lessons from marketing *all* detergent to sell them. Names like Rogue, Lambast, and Limit started to appear at farm stores everywhere. Smart marketing led to soil testing and custom application services at the company's retail fertilizer outlets: why limit yourself to selling something when you could also recommend how much your customers should buy and then apply it for them?

Then along came Randox, a true breakthrough product. Corn and soybeans are usually grown in a two-season rotation—corn one year, soybeans the next. Before Randox, atrazine was the grass control of choice in corn. But it required such high rates and lasted so long in the soil that soybean yields in the following year were compromised. Randox, used in combination with much lower rates of atrazine, changed all that and opened the door to full chemical weed control in corn and soybeans. The economic implications for Monsanto were overwhelming. That the new Agricultural Products Division would eventually become Monsanto's most profitable operation would have been hard to see at the time, but there were signs. The 1963 Annual Report talked proudly of Randox being applied at record rates. The 1964 Annual Report showed that acres treated with Randox were twice what they were in 1963.

Willard Cochrane advising presidential candidate John F. Kennedy at the Democratic National Convention in Los Angeles, 1960.

Cochrane's father, Willard Cochrane Sr., manager of a large cattle and horse breeding ranch near Corona, California, circa 1907.

Cochrane with his mother, Clare Cochrane, in Los Angeles, circa 1928.

Ensign Willard Cochrane with his bride, Mary, on their wedding day, Washington DC, August 23, 1942.

Cochrane as a member of the UN mission to Siam (now Thailand), talking to a Thai farmer, 1948.

A meeting with the minister of agriculture of the Philippine government during the world food crisis of 1973. The minister is the third from the left. The Americans, from left to right, are Dale Hathaway, Cochrane, and Nathan Koffsky.

Cochrane lecturing to an economics class at the
University of Minnesota, circa 1980.

Deere's gamble on producing tractors hit the jackpot in the 1950s. At the beginning of the decade, there were 2 million more farms than there were tractors in the United States. At the end of the decade, there were 1 million more tractors than farms. Looked at another way, the number of tractors on U.S. farms doubled in the fifteen years following World War II. What was going on? Deere's 1963 Annual Report, after announcing record sales for the company, told the story very nicely:

> Although total farm acreage in the United States and Canada has dropped some 15 percent since the mid-1950s, the remaining acreage has been consolidated into larger units requiring increasingly greater use of modern farm machinery. With the dramatic improvement in farming techniques, agricultural production and total farm cash receipts have reached all-time highs. This record income is shared by a smaller number of farms, and the higher cash income per farm has provided farmers with added funds to invest in more productive machinery. It is particularly encouraging that John Deere farm equipment has been capturing a larger share of the market.

The 1950s were not only about *more* tractors, but also about *bigger* tractors. John Deere rolled out its "New Generation of Power" at Deere Day in Dallas during the summer of 1960. Thousands attended the circus and watched the unveiling of a stylish, four-cylinder tractor, festooned in diamonds from Nieman-Marcus. Fireworks and the Al Hirt band kept enthusiasm high well into the evening. John Deere soon realized its dream of becoming the nation's number one supplier of farm equipment. "Farmers are no longer hicks, and neither are we," announced Deere president and CEO William Hewitt. It turned out that many farmers were still "hicks," however, routinely asking for the simpler, two-cylinder tractors Deere had discontinued. Each received a four-page letter from the vice president for marketing assuring them that the change was for the better.

I have often heard Cochrane say that "a professor who is not causing trouble is not doing his job." An exchange with Theodore Schultz in the early 1950s shows Cochrane very much on the job in those days. Professor Schultz, who would later win a Nobel prize, was head of economics at the University of Chicago in 1953. He had come out of nowhere, without so much as attending high school, to earn a Ph.D. from the University of Wisconsin. He soon became head of the Economics Department at Iowa State and built the backwater program into a Midwestern powerhouse. When the infamous "Pamphlet No. 5," authored in his department, had the audacity to favorably compare margarine to butter, the dairy industry came to Schultz's door with lynching on its mind. Schultz led, and resigned during, a legendary fight for academic freedom.

Imagine Schultz's reaction when, on a January morning in 1953, he opened a letter from a University of Minnesota professor, eleven years his junior, by the name of Willard Cochrane. The letter included a draft of a short manuscript with the outrageous title "Professor Schultz Discovers the Weather." The gist of Cochrane's paper was that Schultz's new book *The Economic Organization of Agriculture* was hopelessly shallow in attributing the erratic movements of farm prices to the weather. Had the hapless icon taken time to "read the record of farm technological advance" and, thereby, "develop a rigorous, operational concept of supply," he would have seen that farm prices would be unstable regardless of the weather. The new book was, alas, little more than a "turn down a blind alley."

Schultz responded with a restrained and gracious letter, two pages long, welcoming the criticism. He did what he could to convince Cochrane of what any casual reader of the book would see, namely, the non-weather causes of price instability were discussed at many points in the book. He questioned Cochrane's call for a comprehensive theory of the farm economy but strongly encouraged him to work in that direction. A week later he received two terse paragraphs from Cochrane. Three months after that, "Professor Schultz Discovers the Weather" made its debut in the *Journal of Farm Economics*. It continues to be among the profession's more noteworthy articles.

Schultz is the best known, yet hardly the only, person to have his latest works met by unsolicited appraisals as only Cochrane could write them. Kenneth Boulding must have been especially pleased to open a letter that began like this: "I recently read parts of your new little book, *Principles of Economic Policy*, and it prompted me to write this letter. After reading numerous things that you have written I have reached the following conclusions: (1) in the area of economic theory you are a brilliant analyst, (2) in the area of broad social questions you are a wise man, but (3) in the area of agriculture you are a 'nincompoop.' " The remainder of the three-page letter was exclusively devoted to demonstrating point three.

As it turned out, the liberal Cochrane and the conservative Schultz gradually became something of an academic odd couple. Schultz invited Cochrane to spend a year's sabbatical in Chicago during 1958. That was the same year Cochrane's old friend from the USDA, Harry Trelogan, wrote to tell him he had been nominated to be president of the American Farm Economics Association. The respect for Schultz shown in Cochrane's response is even more remarkable than it might first appear. Cochrane was an ambitious man, and to be elected president of his profession was something he had long dreamed of. After saying how pleased he was to be nominated, he went on to say, "Now one question—who did you cook up to run against me? With one exception, I don't care who he is, except that I am curious. The exception is T. W. Schultz. I would not want to run against him, for I would want him to win. He certainly should be elected President of our Association. Cannot it be arranged?" Trelogan assured Cochrane that Schultz had declined nomination. When Cochrane was elected, he used his presidential address to outline the approach he would soon take in drafting a farm bill for President Kennedy.

Eager students flocked to Professor Cochrane's classes during the 1950s. Through his lectures, textbooks, and professional articles, they learned that with macroeconomics the whole was not always the sum of its parts. This was the central idea in Cochrane's rather irreverent rebuttal to Schultz. Weather was obviously an important factor in determining the year-to-year production by individual farms. Even so, one could not reason that these weather-related variations in production led to the farm economy's notorious price "gyrations," to use a term he was fond of at the time. Good weather in one area balanced out bad weather in another, so the cause of unstable farm prices was to be found elsewhere. His students also learned to be suspicious of free market theory. The assumptions of perfect competition used by free market theorists were universally held to be unrealistic. Conservative economists thought the assumptions "approximated" reality and based their policy ideas on a world-view in which the economy acted "as if" it was perfectly competitive. Cochrane, on the other hand, wrote that perfect competition "remains an abstraction, something that does not exist today and never did exist—the norm is some kind of imperfect competition."[2]

The economic analysis underlying his macro-view of U.S. agriculture and his distrust of free markets was laid out in *Farm Prices: Myth and Reality*. Agriculture, like all other aspects of the U.S. economy, was widely believed to be guided by Adam Smith's invisible hand. Left alone, things would work out for the best. If there was too much product on the market, prices would fall and less would be produced. If there was not enough, prices would rise and more would be produced. This, in simplest form, was what Cochrane termed the "myth of the automatically adjusting agriculture." Even the Depression and its big-time farm aid programs failed to put much of a dent in the myth. At most, some argued that on occasion the government might need to lend a hand to get things back on course. Others refused to grant even that. Then came Cochrane, who questioned the system at its very core where prices determined production.

He began by granting that previous work in support of the price system was true. If the price of corn went up, for example, more would be produced and less would be consumed. If it went down, the opposite would occur. In fact, such behavior seemed reasonable for each individual product farmers were churning out at ever-increasing rates. But it was not true of food taken as a whole. In Cochrane's view, "farmers shift readily between potatoes and sugar beets, corn and hogs, but not between farming and banking or farming and manufacturing." This is an important point; in times of overproduction for most industries, labor and capital would be diverted to other industries and the oversupply would be corrected. Instead, "agriculture represents a water-tight compartment within which there is considerable fluidity, but the connective valve between the agricultural compartment and the rest of the economy works poorly and sometimes not at all." There was no way for system-wide adjustments of the type required for a well-functioning free market to occur. The result, in technical terms, was that the aggregate supply of food production was almost completely inelastic. In other writing, he said the same thing in a more descriptive way by comparing the supply of food to Old Man River: it just went rolling along, regardless of product prices.

Economics is about more than supply—demand plays its role, too. As the income of consumers changed, Cochrane granted they might eat more beans in bad times and more meat in good times. The actual pounds of food consumed per person stays about the same, however: "when the family budget is squeezed, we may drop an insurance policy, put off buying a new car, or go without a deep freeze unit, but we keep on eating." The human stomach requires so much food each day, and that's what it gets, regardless of the price of food relative to the price of other things the consumer might buy. The consequence of this for Cochrane was that "the food needs of the human body, together with food habits, conspire to make the aggregate demand for food highly inelastic."

Now the stage was set for the farm price roller coaster. On the one hand, if the demand for food changed for some reason, the supply could not adjust accordingly. On the other hand, if the amount of food supplied by farmers changed, the price system could not do its work of signaling consumers to change the amount they were eating. Unless, of course, the price changes were extraordinarily large. Any small change in supply or demand would lead to a large change in price unless markets were somehow regulated. In some of his earlier writings, Cochrane seemed most worried that these small but disruptive changes would come from demand, a "restless, unbalancing force in the picture." The drastic swings in consumer income from the Depression to the years immediately following World War II justified his concern. As the 1950s wore on, he became more concerned with the supply side and the effects of technology. New technology could, and routinely did, increase the supply of food without regard for the fact that the demand for food was staying steady. The effect was exactly in line with his theory: farm prices were always being pushed to lower and lower levels.

One question remained: If the inevitable result of adopting technology would be lower prices, why did farmers do it? Cochrane explained this drive to produce with what he called the "agricultural treadmill." If farm prices were falling, and a new technology promising higher production came along, what was a farmer to do? Ignore it, because there was already too much product on the market? Hardly. The first farmers to adopt the technology would produce more, but not so much more as to cause prices to fall even further. They would produce more efficiently and at lower costs, thereby putting a little distance between themselves and the dogs at their heels. So far, so good. But prices would fall to even lower levels as more farmers adopted the technology and output increased significantly. Average farmers had to adopt the new technology, not to get ahead, but just to get back to where they were before "Mr. Early Bird," as Cochrane called him, stepped in. These technologies often required more land to make best use of them, but there was only so much land to go around. "Mr. Laggard," the last to look at the new technology, would be "cannibalized" by his more progressive neighbors, an image strangely inconsistent with that of the hallowed family farmer.[3]

In May of 1960 Mary Conger of Triple C Farms in Iola, Kansas, wrote Cochrane and sent a copy of her article "The Farmer's Side of the Case" that had come out in the April 9 edition of the *Saturday Evening Post*. She was understandably pleased that not only had the article been printed in the *Congressional Record* but that everyone from President Eisenhower to farmers from around the country were writing to congratulate her on the heartrending tale of trying to make a living on a dairy farm. Cochrane wrote back that the article was "excellent" and that he had "never read a better presentation of the case for family farmers in the 1950s." What else could he say about her near-perfect, firsthand account of life on the treadmill?

> When that first slash in farm prices came six years ago, we doubled our milking herd in an effort to increase gross income so that, in turn, we

could meet our fixed charges—such as interest and taxes—and the rising cost of things we must buy. We built a new labor-saving milking parlor—a sort of assembly-line milking system—equipped with a pipeline milker and a bulk refrigeration tank. No expensive labor is wasted carrying milk buckets on our farm. But then came years when our crops were cut by drought, hail and wet weather, and we fell behind on the feed bill for the cattle. In good years we struggled to catch up. We tripled the milking herd. Milk prices declined further. Costs went on up. We were on a treadmill, always running faster just to keep in the same place.

With *Farm Prices: Myth and Reality*, Cochrane drew a sharp line in the sands of farm policy. On one side was the conservative world presided over by Secretary of Agriculture Ezra Benson, a man who believed first in Jesus, second in the free market. This was a world in which less government was better government and free markets were prescribed for whatever might ail the farm economy. On the other side were Cochrane and the liberals. For them, government action was good and necessary. Without public intervention, unstable prices would plague farmers and technology would guarantee surplus production. *Farm Prices: Myth and Reality* was read by many, including a senator from Massachusetts by the name of John F. Kennedy.

Cochrane and the farm movement found common ground on the question of how to save the family farm. The solution lay in higher prices for crops and livestock. Even though liberal economists wanted to bring about these higher prices through government programs while the farm movement preferred direct farmer action to control supply, both agreed on the goal. Nonetheless, by the time Cochrane went to Washington in 1961, agriculture had changed in such a fundamental way that higher prices, no matter how they were achieved, could not in and of themselves have done much for family farms. A look at farm income figures for the time shows why.

In 1929 farm gross income for the United States as a whole was $13.8 billion. Net income, the part farmers kept for themselves, was $6.8 billion. Farmers as a group, therefore, kept forty-nine cents out of every dollar they handled. In 1950 gross and net incomes were both higher, but farmers as a group were still keeping forty-seven cents out of every dollar they handled. These were the times in which the higher prices solution was formed. They were also the times in which higher prices might well have had their intended effect. Then, in a very short period of time, the economics of agriculture changed profoundly. Between 1949 and 1960 gross income went up by $8.8 billion and net farm income went up by $1.2 million. Virtually all of the new money being grossed on farms in the Eisenhower years was spent on new equipment, new seeds, new chemicals, and higher priced land. Farmers were keeping just over a penny on the dollar—not forty-nine cents, not forty-seven cents, one cent.

The second half of this century continued to be one of increasing farm sales soaked up by increasing costs. Decade by decade, billions of dollars left the farm each year and enriched off-farm investors. Farmers never saw as little as a penny on the dollar again, but neither did they see 49 cents. For the billions added to gross farm income from 1986 to 1995, the farmers' share was 7.2 cents on the dollar. In 1995 it took sales of over $210 billion to leave farmers as a group with $34.8 billion. The ratio of two dollars gross to one dollar net had become six to one in a half century. For grain farming in the Midwest, my friend Mike Duffy at Iowa State uses a ratio of ten to one for planning purposes.

All of a sudden, policy prescriptions cooked up prior to 1950 could never work again. Money could not be put into farmers' torn pockets by simply increasing their gross income, be it through higher prices, give-away programs, or anything else. Why, then, didn't Cochrane, his colleagues, and the farm movement leaders see this? One reason that comes to mind immediately is that they all cut their policy teeth in a time when farm numbers weren't falling through the floor and the tidal wave of technology was far offshore. Their ideas did make sense when they were formed. Also, the story was not so clear when looked at from the viewpoint of individual farms. Net farm income per farm was inching upward, but all of the gains were made by cutting the same pie into bigger and bigger pieces. In general, this was and continues to be the method of keeping farm income at acceptable levels: get more people out of farming, let the survivors take bigger pieces of a constant net farm income, and give most of the gains to nonfarmers.

As important as these reasons are, however, a more subtle explanation must be considered. The free market theory of the Chicago school and the macroeconomics of Harvard are presented as being radically different in economics classes. In many ways, they are. In one fundamental way, however, they are identical: both grossly simplify the world into one of producers and consumers. There are people who grow food, and there are people who eat food. Period. John Deere, Monsanto, Pioneer, and Cargill fit into neither theory. Invisible both to theory and to those who developed policies based on those theories, nonfarm corporations thrived while agricultural economists of both schools wrung their hands over low returns in agriculture.

During one unforgettable Tuesday lunch Cochrane and I shared, I asked him why he, the country's best known agricultural economist at the time, heading the entire Economic Research Service under Kennedy, could not tell me anything about corporate profits of the day. "It never occurred to us to look," he said. "You might as well have asked us the diameter of Jupiter."

Professor Cochrane, viewed solely through his many books and professional articles, appears completely taken with the idea of saving the family farm with higher prices for crops and livestock. Privately, however, he was more concerned with the tide of technology sweeping the heartland than he was with low prices per se. There is an occasional hint in his professional writings, such as this passage from *Farm Prices: Myth and Reality*: "the substitution of machines for men may reorganize the family farm out of existence in years to come." But his letters provide a much richer view of a man torn by an idea that would never be popular, no matter what scheme was put forward to advance it. That idea was the control of technology itself.

Reservations about technology, especially technology that put food on the table, were seldom heard in the late 1950s. *Time*'s cover story of March 9, 1959, was much more in tune with the times. Farmer North, presiding over a modern mechanized farm, was described as eating a hearty breakfast before feeding four hundred cows and five hundred pigs in ten minutes. He then returned to his living room, "40 feet long and beige-carpeted wall to wall," for a second cup of coffee while reading a *Wall Street Journal* article on the farm problem. The morning ended with practice on his two-story pipe organ.

The *Time* article left little room to question the new way of doing things. The technological transformation in crop production was duly noted, followed by this comment on the next step toward a "push button cornucopia": "Now farmers are taking the big step from mechanization to automation in the raising of animals and fowl; they are copying the assembly-line techniques of industry and bringing animals indoors. Once man felt he could not provide an environment for animals as good as nature's. Now he knows he can do a whole lot better. Behind him, giving him confidence, are ever-new discoveries in antibiotics, hormones, climate control, nutrition and plant and animal genetics."

When the Farmer North article came out in *Time*, Cochrane wrote John Brewster, a long-time friend at the USDA. He was worried about what would happen to family farms "as the machine technique sweeps over agriculture which I believe is just around the corner." Answering his own question, he saw "new technology literally obliterating agriculture as you and I knew it in our youth." Supply control and higher farm prices could save those farms, but only if those policies were accompanied by some way to control technology. Try as he might, Cochrane could not come up with a clean economic program to do so.

The joy of working with his friends at the Bureau of Agricultural Economics gave way to bitterness and frustration toward many of his university colleagues. He felt his fellow economists were so enthralled with economic growth that the inevitable loss of family farms seemed to them like a small price to pay. Physical scientists in the agricultural colleges were as bad, if not worse. They not only contributed to the problem with their misguided efforts but were clearly in the driver's seat and would never consent to any program that threatened or restricted their research. Even if the colleges could somehow be redirected, the barn door had been left open so long that university-trained scientists were busy doing the same thing, and more, for private industry.

Farmers fared no better in Cochrane's appraisal. They didn't know enough economics to understand how they would make more by producing less and didn't have the common sense to listen to Cochrane when he told them. Instead, they were concerned only with their freedom to do as they wished, regardless of the long-term costs. Only a financial crisis of major proportions would wake them up, and such a crisis had been prevented by government programs throughout the 1950s. Farmers were, as he said in one letter, perfectly willing to accept the "free enterprise pap ladled out in big doses by urban business leaders and their publisher lackeys."

The Cochrane who went to Washington saw himself as a lonely man on a heroic mission to save the family farm. He could expect little support, not from colleagues, not even from farmers. He didn't know exactly how the family farm should be saved, for a way to control technology remained elusive as ever, but he would pursue the supply control doctrine nonetheless. As he set out for Washington, a farm wife in Ohio wrote a long and thoughtful letter describing her lot as "a rebellious liberal in a highly conservative area." In his response, Cochrane told her that "only once in a blue moon do you get a wave of support behind you; the rest of the time you must go it alone and the fight itself is your only reward."

Captain Ahab comes to mind.

4

Professor Cochrane Goes to Washington

In the mid-1950s President Eisenhower and Secretary of Agriculture Ezra Benson found their most ardent support at Purdue University. Various faculty were usually on tour one place or another to defend the free market and a business-as-usual approach to managing agriculture. To enliven their show, they often invited a self-styled "tame liberal" by the name of Willard Cochrane to represent the opposing point of view. He rose to the challenge with such clear thinking and wonderful writing skills that he soon began to steal the show in some circles. The National Farmers Union, one of the largest farmer groups of the time, became particularly supportive of the professor from Minnesota and added much-needed political backing to his policy program. Within a very few years he was a man to be reckoned with, not only in the classroom but in national discussions as well.

Orville Freeman was the governor of populist Minnesota in those years. While lacking any knowledge of agriculture—Cochrane once recalled a long conversation with him about policy options for grain sorghum, only to have Freeman ask, "What is grain sorghum?"—Freeman was a very capable politician. He was also well connected and lived under the wing of Senator Hubert Humphrey. Freeman set up a special governor's commission to study the problems of Minnesota agriculture and asked Cochrane to chair it. This made official what had been brewing for a while—Cochrane was now the leading agricultural advisor to liberal politicians. When Humphrey announced his intentions to run for president in 1960, it was only natural that Cochrane would work closely with him.

The relationship was productive but short-lived. John Kennedy beat Humphrey so badly in the West Virginia primary that Humphrey threw in the towel, leaving Cochrane with nowhere to go but back to the classroom. But the Democrats in general, and Kennedy in particular, didn't want to lose their leading agricultural spokesman. A trip to St. Paul by Sargent Shriver, followed by a phone call from Kennedy, had Cochrane applying for leave from the university and headed for the nominating convention in very short order. He arrived in Los Angeles a day or two before the convention got underway and reported immediately to the Kennedy headquarters at the Biltmore Hotel. After being introduced to several campaign staffers, he was shown his "office," a desk and phone in a hospitality room shared by farmers and their representatives on one side and various civil rights groups on another.

That he was a long way from campus was starting to sink in. He was soon invited to lunch with Bobby Kennedy, Ted Kennedy, and Mike Feldman, Kennedy's right-hand-man in agricultural matters. On the way to lunch, Bobby turned to Cochrane and said rather casually, "I understand that you are going to take care of the farm plank in the platform for us." Without thinking, Cochrane replied, "I guess so." This was decidedly not the response Bobby Kennedy was expecting. He turned and demanded, "Are you or are you not going to take care of the farm plank for us?" A shaken Cochrane quickly revised his response and never forgot his first lesson in the seriousness of presidential politics.

In spite of their rocky start, Bobby Kennedy supported Cochrane when he most needed it. Cochrane weaved his way through the process of selling the Kennedy farm program (which was for the most part pure Cochrane) at platform committee meetings. He confided to Bobby Kennedy that he wasn't sure how to best do this after the first meeting. In return, Kennedy assured him that he was never farther away than a phone call to his secretary. Cochrane called on Kennedy for help, which was always quickly supplied, but generally found an agreeable platform easier to come by than he had originally thought. Democrats of all stripes favored family farms, adequate farm incomes, and feeding the poor and hungry here and abroad. There was also some innocuous language on the need for supply controls, the first mild breezes from a storm far offshore.

Cochrane became acquainted with John F. Kennedy at the convention. The two of them went together to several farm state caucuses. This arrangement worked well, for Kennedy was a self-described ocean and beach person, mystified by the Midwestern farm landscape and longing for a good fish chowder as he dutifully listened to ranchers describing prized cattle. Throughout the remainder of the campaign, Cochrane was there whenever Kennedy came close to farmers. It has recently become fashionable to dredge up the seamier side of American presidents, Kennedy included, but Cochrane will have none of it. Today, as then, he regards the man with the greatest respect and admiration. Kennedy was well educated, immaculately dressed, and always kind in his dealings with Cochrane. He could at the same time make the tough political decisions one sometimes has to make, even if, as in the case of choosing Lyndon Johnson as a running mate, Cochrane could never understand or accept them.

The degree to which Cochrane was taken by Kennedy shows best in a story about a Sunday morning on the campaign trail in Wisconsin. Cochrane was traveling in the candidate's car, briefing him on one aspect of farm policy or another. The car pulled up to a Catholic Church and Kennedy got out. Noticing his advisor's discomfort, Kennedy said, "Come on in. It won't hurt you." Knowing what I do of Cochrane's agnostic views on religion, I suspect it did hurt him. Nonetheless, Kennedy's influence allowed Cochrane to add attendance at a Catholic Mass to his already impressive resume.

The work of advising a presidential candidate was always exciting but not always as glamorous as Cochrane thought it would be. For example, one day he was working in his Washington office, doing the things a political advisor does: responding to letters both friendly and unfriendly, writing brief memos to the candidate on important happenings in agriculture, and working on a policy statement. He got a call from Senator Kennedy's office on the Hill to get over there in a hurry. By the time Cochrane got to the office, Kennedy was ready to leave and was asking for a quick lesson on marketing orders and agreements.

The two men jumped into Kennedy's sporty convertible and were soon buzzing across town toward Kennedy's home in Georgetown. All the while, Cochrane was competing with traffic noise and other distractions as he tried to enlighten the candidate on the subject at hand. When they got to the house, they went immediately upstairs and were met by a barber. The three then repaired to the bathroom. Kennedy pulled down the toilet seat and sat there while the barber trimmed his hair and Professor Cochrane, perched on the edge of the bathtub, continued his lesson. When the barber finished his job, Cochrane and Kennedy walked downstairs while a few last minute questions were answered. Then Kennedy climbed back into his car and drove away. Cochrane, to this day unsure of why the information was needed or how it was used, hailed a cab back to his office on Connecticut Avenue.

Candidate Kennedy and an entourage of three airplanes left Washington DC on September 21, 1960, for an extended campaign through the Midwest and the Rocky Mountain states. There were planned stops in Tennessee, Iowa, South Dakota, Montana, Wyoming, Colorado, and Utah. Kennedy planned to give a major speech on farm policy at the National Plowing Match near Sioux Falls, South Dakota. Cochrane had worked long and hard on the speech that was to be Kennedy's defining statement on farm policy. The speech had gone through numerous versions in Washington and many of the candidate's supporters had added to the version Cochrane carried on the airplane early that morning.

Sometime later in the afternoon of that day Cochrane and Richard Goodwin, Kennedy's number two speech writer, got together to polish the speech Kennedy would soon deliver. Goodwin was to provide the Kennedy prose style while Cochrane stood guard on the substance of the address. The two men worked far into the night and brought the finished product to Kennedy's hotel room at six o'clock on the following morning. Kennedy read the speech sitting up in bed, made a few comments, and requested no further changes. At seven that morning, Kennedy put in a call to Hubert Humphrey and read the entire speech to him. Humphrey, too, approved, and that was it. The speech was ready to go.

Candidate Kennedy made three quick campaign stops in Iowa and South Dakota before arriving at the Plowing Match grounds on the afternoon of September 22. Cochrane and a friend with the *New York Times* watched from the crowd as Kennedy first blasted Nixon and Benson for the farm programs of the 1950s, then moved through the new economics of the Democratic platform. The audience, if not exuberant, was at least friendly and applauded at the right times. Cochrane remembers the moment with great pride. The rest of the day, however, was a different story. In a 1996 interview, he recounted the first of many rude lessons in politics:

> One of the Kennedy lieutenants asked me to ride in the press plane and explain our farm programs to the press. I said that I'd be pleased to do that. They had it worked out so that I was in the back of the plane facing about thirty of the leading reporters from the *New York Times*, the *Washington Post*, the *Los Angeles Times*, and so on.
>
> They started asking questions and I could see right away they weren't interested at all in our marvelous farm program proposal. They had only one interest: What was this program going to do to food prices? It dawned on me that I was suddenly facing a major career choice. Was I going to be a politician, and tell them that it wouldn't do much of anything to food prices, or was I going to remain an economist and give them some realistic estimates?
>
> I knew that being an economist would finish me as a politician and maybe Kennedy as a presidential candidate. I thought for a moment of a

third option, jumping out of the airplane. Maybe I should have taken it, looking back, but instead I tried to compromise and give them what I thought was a very conservative estimate of the food price effects. I said it would raise food prices by about 10 percent.

The next day every major newspaper on the East Coast was carrying a story about how Kennedy's agricultural advisor has said his farm programs would raise food prices by 10 percent. Needless to say, Kennedy's public relations lieutenants called me in and chewed me out and generally raked me over the coals. Kennedy never chewed me out, but there were several anxious days before the story began to die down. I never got out of the doghouse as far as certain key Kennedy advisors were concerned.[1]

Presidential candidate John F. Kennedy liked to say his farm goals were "as common sense as A B C D—for abundance, balance, conservation, and development." His political platform concerning agriculture was laid out in the Plowing Match speech, then refined in the white paper "Agricultural Policy for the New Frontier." That document, again written by Cochrane, used a broad brush to paint a picture of an agriculture brought into balance by supply controls on the one hand and enhanced domestic and foreign food use on the other. Any troubling mechanics of how to do this would be left for the secretary of agriculture to handle after the election.

The family farm was front and center from the very first page of the paper: "We must assure to the American family farm, which produces this abundance, an economic climate in which farmers can earn a fair income—an income which yields farmers a return to their labor, management and capital equal to that earned by similar resources in non-farm employments." The seriousness of the farm problem was addressed in many places, as was the New Frontier goal of stepping in to make things right:

> One of the great issues confronting agriculture and the nation is the economic survival of the family farm pattern of agriculture. The owner-operated family farm, where managerial skills, capital investment and labor are combined in the productive enterprise, is at stake. . . .
>
> Non-farm capital is taking over the managerial function in agriculture, reducing the members of independent farm families to the status of laborers. . . .
>
> The family farm should remain the backbone of American agriculture. We must take positive action to promote and strengthen this form of farm enterprise. This, I believe with all my heart, we should do.

Saving the family farm was important, but so was saving the national budget. The strategy of preserving all family farms with expensive price support programs was, therefore, losing ground to one of saving some at the expense of others. Virtually every agricultural economist of the time thought it was unfortunate, but true, that saving the family farm would require further reductions in the number of farmers so the remaining group could get big enough to survive. How many should leave? The Committee for Economic Development, a group of two hundred conservative businessmen and educators, recommended programs geared toward eliminating one-third of the farm population in a five-year period. After hearing a group of his colleagues present papers on the necessity for fewer farms at the 1963 meetings of the American Farm Economics Association, Cochrane's colleague at the University of Minnesota, Phil Raup, was reported as saying: "We applaud men who can create new jobs in industry or service trades, but agricultural economists seem to devote their professional efforts to devising new ways to destroy jobs in agriculture."

The general story of swearing allegiance to the family farm, then lamenting the fact that its preservation would require some farmers to leave so the best farmers could get larger, proved to be remarkably enduring. In 1997 the CEO of Novartis Seeds, one of the largest multinational suppliers of seed and technology, justified the move to what he called "agricultural industrial complexes" on, what else, "the movement to fewer but vastly more sophisticated—and demanding—farmers."[2] Meanwhile, barely 120,000 farms were producing 60 percent of the nation's food supply.

In 1960 Kennedy was elected president, Humphrey was sent back to the Senate, and Orville Freeman lost the governorship of Minnesota to the Republicans in a very close race. The usual story is that he lost because his years of free-spending liberal programs finally caught up with him and he had to raise taxes. Privately he blamed his loss on being chosen to make the nominating speech for Kennedy at the Democratic convention. Endorsing a Catholic for president didn't sell well at all in Lutheran Minnesota. In any event, Freeman expected that the New Frontier would mean his period of unemployment would be short-lived. High posts in the attorney general's office, the Department of Interior, and the Department of Defense were bantered about. Freeman was agreeable to any position in the new administration with one exception. He was widely quoted as pleading, "Don't let them make me secretary of agriculture."

Freeman didn't understand farming, but you didn't have to in those days to know the Department of Agriculture was a fabulously expensive bureaucratic quagmire. The story was everywhere. *Life, Saturday Evening Post, Time, Reader's Digest*—they were all full of stories about what the previous secretary, Ezra Benson, had referred to as a "monster" and a "sordid mess." The population was much more urban than rural in 1960, so barbershop analysis of the Department of Agriculture was slanted more to its cost than to saving the family farm. How much was the department spending? The figure $7 billion per year doesn't fully capture it. Perhaps a better way to see it would be to add the combined budgets of the Commerce, Interior, Justice, Labor, and State Departments. Then multiply that number by two. Then add some more.

For roughly twenty cents on every non-defense tax dollar they paid, what were the good citizens getting for their largesse? To read contemporary commentary, not much. Mountains and mountains of surplus farm products had accumulated from past years. No one knew what to do with them, so the department spent a billion dollars or so each year for storage until some better idea came along. Many more billions were spent on programs to reduce the amount farmers were producing. Some farmers were bought off the land entirely with the Soil Bank Program, but most agreed to reduce the land they planted in exchange for price supports. So it wouldn't look so much like a handout, farmers were given loans when they harvested their crops and put them in storage. If the crop was worth less than the loan, the farmer simply walked away and left the public holding the crop.

The program, ineffective as it was, took an army of bureaucrats to administer. USDA employees, over one hundred thousand strong, nested in Washington, in every county in the United States, and in several foreign countries. As a bit of a lark, a congressman introduced a bill in 1962 requiring that the number of USDA employees "shall at no time exceed the number of farmers in America." His colleagues in the House took it as a serious gesture and came within thirty votes of passing it. Scandal, perhaps inevitable in such a large organization, brought more cries of outrage. Someone was discovered printing counterfeit money in one of the forty-eight hundred rooms the department occupied in Washington. Billy Sol Estes made the cover of *Time* for his fraudulent escapades in the cotton program. Mostly, however, the stories were of very wealthy farmers getting huge payments from taxpayers who would never earn as much in their entire lives. Finally, and most difficult to explain, a good many USDA scientists worked at things like breeding higher yielding varieties and discovering new ways to combat insects. Helping farmers learn how to produce more seemed like an unusual response to expensive crop surpluses.

Freeman fidgeted as he was passed over for one good job, then another. Finally, when all were gone, President Kennedy called and offered the agriculture post. Freeman accepted, then asked Cochrane to serve with him. The political pendulum at the USDA had swung from far right to far left in a matter of months.

Freeman and Cochrane faced an enormous task. In addition to the obvious troubles facing the agricultural economy in general, the USDA itself was not set up to deal with those problems in a way consistent with the Democratic program. The first order of business, then, was to establish a USDA in which economists could effectively deal with the policy problems facing the nation. In short, they would reestablish the old Bureau of Agricultural Economics.

The bureau had been hounded throughout the 1940s and early 1950s by a group of conservative Southern congressmen opposed to its liberal views. Jamie Whitten of Mississippi was among those especially outraged by a bureau report on the deplorable economic conditions facing African Americans in the rural south. Howard Tolley, chief of the bureau, was forced to resign in the late 1940s as the economic noose around the agency was steadily tightened. Finally, when the Republicans gained power in 1953, the bureau was dissolved altogether and its economists and statisticians were scattered among other USDA agencies such as the Agricultural Research Service, the Agricultural Marketing Service, and the Foreign Agricultural Service.

This represented a fundamental change in the role of economists in agriculture. Broadly speaking, economists are concerned with questions of how best to govern agriculture when the political winds are blowing from the liberal camp. During conservative times, however, there is no question of how to govern the agricultural economy—the free market will do nicely. Hence, economists tend to serve other disciplines, agronomy and animal science, for example, and do such things as help farmers decide if adopting this or that technology will be to their benefit. The sum of these well-informed, businesslike individual decisions is taken to be the best of all possible worlds.

The dismemberment of the Bureau of Agricultural Economics did not go unopposed. In particular, John Kenneth Galbraith spearheaded a letter writing campaign among economists across the country. The effort did not succeed, but Galbraith never gave up. In 1960, as one of John Kennedy's most trusted advisors, he convinced the candidate that he should once again establish the bureau or something like it were he elected president. Cochrane, of course, agreed with this wholeheartedly and threw himself into the task as soon as Freeman took over.

Harking back to his unfortunate times in the Bethesda Hospital, Cochrane once described the task of prying economists out of their latest agency homes as akin to extracting imbedded wisdom teeth. Mountains of paperwork were involved: Civil Service Actions to be taken, justifications in support of new agencies, and directives written for the secretary's signature initiating actions to transfer personnel and budgets. In the remarkably short period of two months, the modern-day Economic Research Service and Statistical Reporting Service were established under the leadership of their first director of agricultural economics, Willard W. Cochrane.

The new organization had one major shortcoming—it had no permanent funding. If it was to exist beyond the fiscal year, it would require a budget of its own. Such a budget would need to clear the Agricultural Subcommittee of the House Appropriations Committee. This committee, as it turned out, was chaired by none other than Jamie Whitten, by now such a powerful man that he was known about town as the "permanent secretary of agriculture." Cochrane was designated to present the case to Whitten for two reasons. First, as the one who had designed and set up the new structure, he was a logical choice. Second, no one else was willing to do it.

Cochrane began his meeting with Whitten by explaining what they had done, even though the well-connected congressman was way ahead of him on that one. He presented the new organizational chart and went through the list of who would fill each of the key positions. Then, in a timorous voice, Cochrane asked for Whitten's approval. Whitten began as Cochrane expected he would: he was opposed to the new system and putting so many economists in one place would only lead to trouble. But then he went on to say that the president wants it, and what the president wants, the president shall have. Once again, Cochrane was in awe of Kennedy's political powers.

For all practical purposes, the new organization was now a reality. But Whitten was not quite through with Cochrane. He reminded him of what he called "hound dog studies," those from the old Bureau of Agricultural Economics dealing with African Americans in the rural South. Should the new agency produce any more such reports, there would be serious trouble. Cochrane assured Whitten that there would be none. When he left Washington in 1964, Cochrane paid Whitten a final visit. Whitten took the occasion to acknowledge Cochrane's effective management of the new organization and to thank him for keeping his word on the hound dog studies.

The most enduring legacy of Cochrane's years in Washington had little to do with traditional commodity programs. Within a few months of taking office, he and Freeman established the Food Stamp Program in its modern form. Cochrane's interest in domestic food and nutrition problems began with his Ph.D. research at Harvard. He published USDA Bulletin 581, *High Level Food Consumption in the United States*, soon after he graduated and continued his food consumption research during the 1950s. Long discussions with Freeman during the time he was governor of Minnesota, then with John Kennedy during the presidential campaign, greased the skids for fast action on a new approach to domestic food aid when Kennedy took office.

Prior to 1961 food aid to the poor was a matter of the government buying up surplus farm products and giving them away through school lunch programs, direct relief, and the like. The poor had little say over what they received under this scheme. If raisins were in surplus supply, the poor got raisins. Cheese and powdered milk were perennial favorites. Food relief, to the extent it was part of USDA, had as its primary goal the disposal of surplus farm products, not of providing interesting diets for the poor.

The Food Stamp Program was an attempt to preserve the USDA goal of using up surplus production while at the same time providing more freedom among the poor to choose what they ate. The poor would be given coupons that could be exchanged for whatever food they wanted at ordinary stores and supermarkets. Here the twin goals of helping the poor and disposing of surplus food came into conflict. If you gave people food coupons, what about the money they were already spending on food? Would they use that to buy still more food, or use it to buy other things now that their food needs were covered by coupons? It was widely held that the latter option would prevail.

The original food stamp idea remained true to the surplus disposal goal. A poor family would work with an agency to determine their "normal food expenditure" for some period of time. Say that amount was fifty dollars. Under the program, the poor family would give USDA fifty dollars in cash and get one hundred dollars worth of food coupons. They would have no choice but to double their food consumption under this plan. Two objections killed the idea. For one, by the time the poor were identified and brought into the program, they often had no cash to buy coupons. Second, the Agricultural Marketing Service, the branch of USDA charged with administering federal food programs, wanted nothing to do with the daunting process of determining the normal food expenditure for each and every family in the program.

The compromise brought forward in 1961 was a plan of simply issuing food stamps at no cost to the poor. Cochrane knew that no program for the poor could solve the surplus problem, especially a program that allowed recipients, in a roundabout way, to spend their food stamps on things other than food. The program was much more about social welfare than anything directly within the scope of USDA. Cochrane, always the liberal, found the compromise acceptable for other reasons: the government was helping people who needed help, and money was being taken from the rich and given to the poor. There were also important political implications for his hopes of advancing programs to slow down the production of surplus farm products. Liberals could not be expected to support limiting food production as long as there were people not getting enough to eat. The sales pitch, then, was to be that the food needs of the poor would first be provided for. Any remaining surplus production would be addressed with supply control.

Shortly after Kennedy took office, Freeman and Cochrane discovered they had the legal authorization to implement an experimental food stamp plan. Thus, one of their first actions was to put a limited plan in place for several cities throughout the United States. The program grew steadily during the 1960s, then mushroomed during the Nixon administration. The Food Stamp Program, with an annual budget on the order of $25 billion in the late 1990s, is now the largest single program administered by USDA.

Now it was time to face the problems that had brought the Democrats to Washington: surplus production, costly government programs, and low product prices for farmers. A 1959 issue of *Life*, featuring a cover story on Marilyn Monroe's latest movie, carried an editorial that was common in both substance and style. Readers were advised to "start squawking loud and long enough to put an end, for good and all, to this incredible Farm Scandal that is getting worse every year." University economists, on the other hand, were downright apocalyptic in their assessment of a return to free markets. Farm income declines in the 50 percent range, farm prices falling into the basement, and unprecedented losses in farm numbers were predicted. These losses, in turn, would roll through rural America with the force of a flooding river, finally affecting the U.S. economy as a whole.

A simple, though draconian, way out of the surplus problem was simply to dump everything in the ocean and be done with it. Early on, Cochrane had his staff evaluate a variation on this theme. At then-current projections, each bushel of surplus grain would cost more to store than it could ever bring when sold. Destroying the surplus farm products, sensible as its costs might be, would, of course, set liberals howling and was not seriously considered. Still, it underscores the magnitude of the 1960s farm policy dilemma.

Cochrane also rejected a business as usual approach to farm policy. In his view, farmers were living in what he called a "fool's paradise" of public handouts that could not continue for much longer. He was dead wrong on this account, even though his reasons made sense at the time. For one thing, he was convinced that the public would soon revolt over the high costs for farm programs. In fact, twenty-five years later program costs dwarfed anything that could have been imagined in 1960. For another, he had no confidence that exports at realistic prices would ever amount to much. Within fifteen years, the Russian wheat deal would have a new secretary of agriculture urging farmers to "plant fence row to fence row." Worst of all, he thought the public would see higher food prices as a small price to pay for a program that would control production without costing so much in taxes.

His solution was outlined in an article called "Some Further Reflections on Supply Control." Soon to be dubbed "The Cochrane Concoction" by its legion of detractors, it was a grand scheme to take the overproduction bull by the horns once and for all. On the face of it, the plan was radical but otherwise simple enough to understand. Each year, a government agency would estimate what the demand for each agricultural commodity was likely to be at a fair price to be determined by Congress. Then farmers would be allowed to sell only that much.

Cochrane reasoned that his plan was not that much different from programs that had been in place for decades. Those programs tried to control output by controlling a single input: land. If farmers had less land, they would obviously not be able to produce as much. Unless, of course, there was a constant flow of new technology that offset the land controls. That was exactly what had been happening and what was likely to happen for the foreseeable future. Cochrane toyed with the idea of somehow restricting technology in agriculture and promised a new paper in his supply control article that would outline how such a major redirection of agriculture might be accomplished. But he also acknowledged that progress had become something of a new religion in America and that the program would be virtually impossible to sell. As it turns out, he never wrote the technology paper and put all his eggs in the supply control basket. He proposed each farmer be issued a marketing certificate, or quota, that would allow him or her to sell a certain amount of product. How to produce that amount was left entirely up to the farmer, but production above that amount would not be allowed. Some began to call the idea franchise farming, which is not such a bad analogy.

Academic economists are bred to believe that people are basically rational and will soon enough see the world as any right-thinking economist sees it. Cochrane was very much of this tradition. At first blush, his program might look like it restricted farmers' freedom. Once they gave it a chance, however, farmers would see that it was in their best interest to support the program. As an added bonus, farm prices would be liberated from the capricious tyranny of the free market. Prices would be high enough for farmers to make a good living and stable enough for them to plan for their next year with a confidence they had never before enjoyed. Consumers might see somewhat higher food prices, but they would be assured that they were fair and determined by their elected officials. The higher food price pill would be made palatable because taxes would go down. No longer would the taxpayer need to pay farmers not to produce.

Cochrane's ideas may have gone without question in college seminars. In the real world of politics, however, there was the Cold War to consider. Communists were thought to be lurking around every corner, waiting to do their dirty work if one crusading senator or another didn't send them packing back to the Kremlin where they belonged. To many minds, the danger of the supply control idea was that it paralleled what happened in Mother Russia. Popular commentary often took this simple approach:

1. Cochrane is a Communist.
2. His ideas are fit only for Communists.
3. We're Americans.

From every front, in never-ending sarcasm and slander, Cochrane and his ideas were held up for ridicule and gross misrepresentation. Here are but a few excerpts from a thick file of Cochrane's labeled "Publicity: Killer of Freedom":

> Unfortunately, the plan, as laid out, smacks too much of things which are everyday occurrences in Communist countries. We don't have to ape the Russians and Chinese, for if that happens we might as well change the title of Irving Berlin's inspiring song to "God Help America." (*The Northwestern Miller*)

> Having ridden into Minnesota's governorship with the "arranging" of James G. Patton's socialistic National Farmers Union, Freeman now has his USDA bailiwick "packed" with NFU influence through such detrimental characters as the University of Minnesota's Dr. Willard Cochrane and others of the same detested communistic stripe. (*Grain and Feed Journals*)

In another issue, *Grain and Feed Journals* helped its readers interpret the "gobbledygook propaganda from this camp of Khrushchev followers": "'Supply management' is a sugar-coated expression cleverly substituted for a dangerously realistic description of a ruinous approach by our Washington agricultural commies for their program."

Of all those laboring to protect unsuspecting Americans from Cochrane and his gang of Commies, the most zealous by far was Bill Kennedy, editor of *Farm and Ranch*, "serving 1,140,000 subscriber families in the sunbelt of agriculture, coast to coast." He wielded his "Washington Wire" without mercy and penned such headlines as "Cochrane Concoction: Killer of Farm Freedom" and "Time *Now* to Clobber the Cochrane Gang." Kennedy was apparently well connected and was able to add injury to insult by having such gems as reference to "a 'mad economist' named Willard Cochrane" placed in the *Congressional Record*. He also provided key congressmen with advance copies of his columns, which included solemn assurances that "socialized farming" was even worse than "socialized medicine, Federal 'take charge' of public education, and sharp increases in Minimum Wage" that were other "pieces of legislation violating the American concept of government."

Cochrane's "Killer of Freedom" file is one few could amass and none would want. How do you react when you come into work one day and find still another published story misrepresenting your ideas? Worse yet, it is given to you by the secretary of agriculture, who in turn got it from an angry farmer who scrawled in the margin: "As a taxpayer, I object to keeping this man in my employ. Mr. K. could use him." The Communist charges against him were so infuriating that Cochrane had USDA lawyers prepare background for a lawsuit. The lawyers determined Cochrane did, indeed, have grounds to sue: "the imputation of Communism and in many instances the imputation of mere Communist sympathy is libelous *per se*." More to the point, however, neither President Kennedy nor Secretary of Agriculture Freeman would be too happy to have the head of the USDA's new Economic Research Service involved in a protracted and highly public court battle, with the farm press, no less, over whether he was a Communist, a sympathizer, or merely a left-wing academic.

There is a poignant footnote to the anti-Cochrane fiasco that swept the farm press. After the smoke cleared, a writer for one of the industry magazines that had most tormented Cochrane called him to have lunch. Cochrane accepted and listened patiently as the writer told him how much he admired Cochrane's intellect and how badly he felt over the things he had written. He was just doing his job, you understand.

The farm movement, best represented in this story by the National Farmers Union, was influential in putting Cochrane in Washington. He was sometimes referred to in the press as the "patron saint" of the movement and added much-needed academic respectability to their position. The farm movement was distinct in one important way in regard to the "government action versus free market" debate of the time. Most economists favored some sort of free market solution to farm problems; Cochrane argued for government action. The farm movement, however, had a third solution. They favored direct action by farmers. This sometimes had dramatic presentation, such as milk dumping parties and the killing of baby pigs. In the Kennedy farm bill, the influence of farmers was to be more subtle and far-reaching.

Cochrane had long thought that farmers did not understand economics well enough to draft farm legislation. At the same time, their support for supply control was essential. Congress, too, had problems in that Cochrane thought the long road from a farm bill's introduction to its final approval left too many opportunities for ideas to be watered down and influenced by big business. Nonetheless, no bill could reach the president's desk without congressional approval. The delicate task of obtaining farmer and legislative support for supply control was to come from an innovative process Cochrane drafted for Kennedy's first farm bill. Here, broadly speaking, is how the process he envisioned would work. When farmers growing a particular commodity came to the USDA complaining about low incomes, the agency would set up a farmer committee with which USDA program people could test ideas. The USDA would then come up with a program that would keep supply of the commodity at levels consistent with prices Congress had determined to be "fair." If the farmer committee approved the plan, it then would be sent to Congress. Congress could either accept or reject the plan, but lawmakers could not amend it. A plan accepted by Congress would have one final hurdle, two-thirds approval by all farmers growing the commodity. If such a referendum was successful, the control plan became law and would be mandatory for all farmers.

Secretary Freeman soon endured extended grillings in the House and Senate. Over and over, he told puzzled members of the agricultural committees that the new farm bill was not so much a program as a process. This was no small change from how farm programs had been developed up to that time. Congress was in the driver's seat and wanted to stay there. Some members of Congress enjoyed the enormous power granted them to fashion farm programs. Others feared that the new farm bill gave the executive branch too much power. Congress could approve or disapprove a farmer/agency program proposal but could not modify it. A humorous moment during the 1961 House hearings underscored the problem Congress had with the bill. Congressman Findley led Secretary Freeman through a series of questions that had him admitting that farm policy should be made by people of great knowledge and experience. Then Mr. Findley asked, "Well then, I think you have overlooked a real excellent source of members for your advisory committees. Did you ever consider selecting all of your advisory committees from the Members of Congress?" Secretary Freeman bluntly answered, "Frankly, no."

You would think, on the other hand, that farmers would embrace the new idea wholeheartedly. Unfortunately, much of the debate centered on the fact that any program approved by Congress and farmer referendum would be mandatory for all farmers. Otherwise, the secretary reasoned, the programs would not be effective. Farmers, it turned out, preferred the benefits the old system had given them to the prospect of government control of their markets. Ironically, the issue of "farmer freedom" became a major stumbling block for what would have been the most farmer-led farm program of the twentieth century. There were other problems, too, mostly centered around attempts to explain the difference between a program that everyone understood and a process that no one understood or trusted. Time after time, the secretary was asked if his bill would do this for wheat or that for cattle. His answer almost always was a variation of "I don't know." He couldn't answer otherwise, of course, because those were the very questions his farmer process would answer once the bill was authorized by Congress. Still, such responses did little to build confidence on the Hill for the new secretary or to allay fears from various commodity groups over what the program might mean to them.

Cochrane's process for involving farmers more and Congress less in the policy making process died in committee and never resurfaced. There was still much to learn about the ways of Washington.

The Kennedy administration had been, if nothing else, persistent in its effort to push supply control through Congress. The first attempt never made it out of committee in either the Senate or the House. Stopgap emergency programs were enacted instead. The next year they were back again with the Food and Agricultural Act of 1962. Cochrane was again instrumental in drafting the ambitious legislation. Granted, the farmer process was gone, but tough control programs for dairy, feed grains, and wheat were included. Only a shadow of the original draft survived the congressional gauntlet and became law. The production controls for dairy were dropped altogether and voluntary controls for feed grains from past years were continued. The mandatory program for wheat survived with the condition that it be approved by a farmer referendum to be held in 1963. The stage was thereby set for a dramatic last hurrah of Cochrane-style supply control policies. It is difficult to express how seriously Freeman and company regarded the wheat referendum or how costly its defeat would become. President Kennedy himself went out on a limb, "public service announcements" in favor of the program were on TV everywhere, and over 5 million pamphlets supporting the program were mailed out by the USDA.

The government presented its argument as one of two-dollars-per-bushel wheat versus one-dollar-per-bushel wheat. Under current programs, wheat was supported at two dollars. Without programs, however, wheat prices would be cut in half. The programs would be withdrawn either way, so what did the farmers want? A program of their own to control production and keep prices high or to be cast into the darkness and gnashing of teeth of the market? Those opposed to the wheat control program appealed to the long tradition of independence among American farmers. The government had gone too far, and it was time to rein it in. What became most troubling for Freeman and his staff was that as things got more and more out of hand, the vote looked more and more like a referendum, not on wheat controls per se but on the liberal approach to farm policy altogether.

May 21, 1963, was the last day of serious mandatory supply control consideration for grain in the twentieth century. Not a single wheat state came through with enough votes to carry the plan, and other states chimed in with equally dismal returns. It was not only a defeat for the liberals but a shot in the arm for the conservative American Farm Bureau Federation that had led the opposition. The Farm Bureau played its hand as a "freedom to farm" issue, but there was more to the story. Shortly before the vote, Charles Shuman, their president, addressed the group's national convention in Atlanta. While stopping short of previous statements that the federal farm program "denies the unmistakable pattern of God's law," he otherwise conducted himself well with the obligatory flag waving, condemning of bureaucrats, and appeals to freedom in urging negative votes from the enormous membership. Then the influential "Charlie," as he was known to many in Congress, told the members the real reason to vote "no." There was an election year coming up, and few in Congress would have the guts to follow through on the threat of letting their beloved free market drive wheat prices down to one dollar. Shuman was right, of course—the referendum was barely over when Congress went to work on new wheat support programs featuring federal payments and voluntary acreage controls.

In 1965 *Time* magazine looked back on "the defeat of the 1963 nationwide wheat referendum, which shook the Kennedy Administration to its socks. It was the Gettysburg of the war between farmer and bureaucrat—and Shuman was its General Meade. The referendum's Robert E. Lee was Willard Cochrane, then Freeman's director of agricultural economics, a tough-minded theoretician whose ideas proved politically unacceptable."

The defeat of the wheat referendum led quickly to rumors of a shakeup at the USDA. Cochrane knew the game was over for the type of policy he wanted and started looking for greener pastures. Of his many options, an aggressive recruitment effort by the University of Minnesota seemed most attractive. In the summer of 1964 he announced his resignation and packed his books and papers. A good part of the attraction of going back to Minnesota was an assurance that he would be appointed to one of the first prestigious Regents' Professorships at the University. This would have been a worthy recognition for a man of such stature, but it was not to be. The hounds of conservative agriculture soon tracked him to St. Paul and convinced the administration that to honor a man as unpopular as Cochrane would have grave consequences for university fundraising. The regents' offer faded into the political darkness.

He was trapped both intellectually and financially. Barely able to support his family in Washington, he exhausted his remaining resources in going back to Minnesota. Further moves were out of the question. Instead, he returned to the same classrooms at the same rank as when he had left in 1960. Only now, he was not the determined Ahab who would track the whale at any cost. Now he was Ishmael, clinging to the coffin of his grandest ideas on the gray sea of academia.

On top of everything he blamed himself for the supply control failure. His accomplishments would have been more than enough for anyone else. He had reorganized USDA agricultural economics into a large and effective unit. As director of agricultural economics, he had been a key advisor on important policy decisions, drafted legislation, reinvented the Food Stamp Program, and written speeches for the president of the United States. But he could not forgive himself for misunderstanding the ways of Washington. Had he been a better politician, he felt he would have better served President Kennedy and been successful in implementing the supply control ideas he still felt were sound.

The truth was that no amount of political polish could have brought his program forward in such an economically hostile time. Agriculture was already well down the road it travels today: "Monsanto again benefited from the healthy American agricultural economy. Meanwhile, there were welcome signs that the emphasis of U.S. farm policy was starting to shift from surplus control to increased production. Moves in this direction tend to swell demands for farm chemicals."[3]

5

An Unreconstructed Liberal

On November 15, 1961, the *Washington Post* carried a story with the headline "Surplus Held Unsuited to World Needs." The main point of the article was that, according to "the Administration's most influential farm economist," the surplus production of U.S. farmers was in large part not suitable for feeding the world's poor. Cochrane was quoted as saying that the world's food needs "would be better served, for example, by producing less feed grains and more soybeans and dried peas and beans." Cochrane did not say this because he thought exports would ever do much to solve the surplus production problem. Rather, in the best liberal tradition, he wanted to find better ways to feed the world's poor and use American food aid as a means of improving the economies of developing countries.

For other liberals, including President Kennedy, surplus food was an important part of foreign policy. Kennedy, a big proponent of "Food for Peace," was convinced that the only thing Russian leader Kruschev truly feared was being unable to feed his people. When Secretary Freeman presented the Agricultural Act of 1961 to Congress, he acknowledged that food aid was a longstanding tradition in America. "What is new," he explained, "is the magnitude and supreme importance of the task today." After allowing that expanded food aid programs were "the human, the generous, the American, and the right thing to do," he went on to remind the legislators present that such actions were also "essential to our own national interest and national security." By this, he meant that properly targeted food aid could be useful in persuading emerging nations to adopt the ways of democracy rather than those of Communism. In its food abundance, the United States had "a weapon of unquestioned superiority."

Farm policy in Camelot held that preserving the family farm structure of

American agriculture was an essential part of feeding the world. Several examples of how important family farms were to Cochrane and Freeman have already been given. Support for family farms was much broader, however, and permeated the liberal thinking of the country. For example, shortly after the defeat of the wheat referendum, the *Washington Post* carried an editorial stating that: "it is an unexplained fact of history, but an undoubted fact despite its mystery, that great nations have always tended to decline when independent farmers were driven out by industrial farming."[1] Freeman, again in his testimony on the 1961 Farm Bill, made it clear that family farms were at the foundation of the liberal farm agenda. "If low incomes squeeze out all but a few corporate-type farms, there would doubtless result the kind of supply control that would result in higher prices, without regard for the public interest, or the consumer interest, or interest in our programs to expand the use of food abroad in the interest of peace and human progress."

Our farm policy at that time was intended to support the goals of protecting family farms, feeding the hungry here and abroad, and advancing U.S. foreign policy. The principal means for advancing these goals was a rather complex price support program. Through this program the government owned surplus grain but did not generally own storage facilities or ocean vessels. Because of this, private grain companies would have to be induced to cooperate in liberal programs they did not support. The delicate task of brokering deals between the U.S. government and private companies involved in storing and shipping grain was handled by a small group of top USDA officials called the Commodity Credit Corporation, or CCC. Cochrane belonged to this powerful group that was, in effect, the Bank of USDA.

The CCC met once or twice a week in Washington to decide the "details" of administering programs that Congress had authorized. Such details included setting program loan rates, export subsidies and acreage set-asides and buying up surplus eggs and such for the school lunch program. Reynold Dahl, a colleague of Cochrane's at the University of Minnesota, remembers him complaining of the nonstop chorus of "gimme, gimme, gimme" being sung by private grain interests at virtually every CCC meeting. In some of these meetings the fight would be over who would get lucrative contracts to store government grain. In others, those with contracts would ask for higher storage payments. When storage rates weren't on the docket, export subsidies took over. For example, say that the government was supporting the price of wheat at $2.00 per bushel when wheat was selling on the world market at $1.30. Clearly, a grain company would not do well by buying wheat directly from farmers and then selling it for $0.70 less. Instead, the grain companies went to the CCC and negotiated an export subsidy that was sometimes paid in cash and sometimes in additional grain the companies could sell.

The liberal farm policy machine was complex and expensive. No one denied that. At the same time it was designed, for better or for worse, to advance liberal causes in ways the free market could not reasonably be expected to do. With free markets, one turns things over to self-interested profit seekers and waits to see what will happen. The outcome of this process, aside from claims of it being the best of all possible worlds, is difficult to forecast. Even the most ardent free market supporter would have trouble saying that such a system would necessarily preserve family farms, feed the world's poor, or advance the goals of U.S. foreign policy.

It is too easy to say that these values fell victim to the farmer vote on the wheat referendum. The wheat vote may have finished Cochrane-style supply control, but the grander ideals for U.S. farm policy were done in elsewhere. The government, embarrassed and burdened by expensive farm programs, struck a Faustian bargain with the global corporations. The soul of liberalism was exchanged for an empty promise of lower public costs to support agriculture. The terms of the deal were worked out over a long time, beginning while Cochrane was still with Kennedy and ending thirty-five years later with a farm bill called, with what can only be termed the cruelest of irony, "Freedom to Farm." During these years Cochrane first faced the challenge of recovering emotionally from what he viewed as his personal defeat in Washington, then the daunting task of renewing his search for ways to advance the Jeffersonian dream in an ever more hostile world.

A month after he was quoted in the *Post* article on feeding the world's hungry, Cochrane received a memo from Sherwood Berg, his department head back at the University of Minnesota. The memo warned Cochrane that he should watch out for "a very shrewd opportunist" by the name of Dwayne Andreas. Everyone familiar with big league agribusiness knows Andreas as the modern-day mastermind of Archer-Daniels-Midland, the self-proclaimed "supermarket to the world." In the late 1950s Andreas was a grain magnate and political operator known well by Cochrane because of their mutual connections with Hubert Humphrey.

To fully understand Berg's warning, we must go back to the National Farmers Union (NFU) in the 1930s. The NFU viewed the Minneapolis Grain Exchange as a private club in which the deck was stacked against farmers. To protect their interests, the NFU supported the establishment of Grain Terminal Association (GTA), a marketing cooperative. GTA and NFU saw eye-to-eye throughout the 1940s and 1950s. When Cochrane published *Farm Prices: Myth and Reality*, the Bible of NFU-style supply control, GTA both contributed money to produce the book and bought hundreds of copies for distribution to key people around the country.

All was not well, however, for the child was becoming wealthier than the parent. The influence of farmers was declining; the action was more in storing and trading the surplus grain pouring from the technological cornucopia. This GTA did well and wanted to do better. To this end GTA purchased Honeymeade Products, then the world's largest private soybean processor. Andreas, Honeymeade's fortyish owner and manager, came to GTA as executive vice president in the deal. Tension began to build as NFU president Jim Patton, Secretary of Agriculture Freeman, and USDA chief economist Cochrane doggedly held to the family farm program of supply control and high prices. Andreas grew up in the grain trade. From his point of view, GTA would fare better in a world in which there was more, not less, grain to store and trade.

Andreas knew politics. He was one of Hubert Humphrey's largest financial backers and understood the strengths and weaknesses of the liberal position better than most. The liberals were vulnerable on supply control in two important ways, both of which Andreas soon was to use to his advantage. For one, liberals were generally concerned about feeding the world's starving people. Reconciling this with cutting food production in the United States was a tough sell. Second, liberals were always vulnerable to charges of spending too much money and favoring too much government. How exactly do you defend an enormous public agency with a budget measured in the billions to limit food production?

While Andreas was positioning himself to undermine the supply control movement, he ran afoul of Cochrane on a different matter. GTA was heavily involved in storing government grain. Grain storage cost taxpayers as much as $1 billion per year in those days, so this was no small matter. Andreas visited Cochrane regularly and one day asked a rather substantial favor. He wanted CCC to pay a higher rate for the grain GTA was storing. As a member of CCC, Cochrane's support was essential. Cochrane's staff analysis showed that rates were already high enough, and pressure to control government program costs was building. Cochrane voted "no." The two men never spoke to each other again.

Andreas made his move in a December 1961 speech at the GTA annual meeting in St. Paul. He cautioned that "we must not allow the luxury of crop-cutting at the expense of taxpayers to go so far that it erodes our markets and eventually turns our hard-won customers over to the Australians, Canadians, and eventually to the Russians." Adding the Russians was a nice touch, embellished with the suggestion we should open trade with Red China before someone else beat us to it. To do otherwise would "make our great nation a laughing stock before the world of merchants." To say the least, Cochrane and company were caught off guard. Maintaining family farms with supply control was giving way to euphoria over staying ahead of the Russians, not only in the space race but in the grain race as well.

Berg ended his 1961 memo to Cochrane with "Once again, everything isn't always economics; there is some fast-stepping politics in our beloved Agribusiness." The story was not over yet, however, for Cargill and its cousins were about to make their move.

Private grain companies, unfettered by GTA's liberal ancestry, wanted the government out of the grain business altogether. Erv Kelm, a man whose uncanny instincts for trading grain had brought him up through the ranks at Cargill, had just taken over as chief executive officer when President Kennedy was elected. He pulled no punches in stating the Cargill position on U.S. farm policy: "I hope the administration will not decide to impose additional controls on farm production or on the ability of firms such as Cargill to handle farm products."[2] Cargill wanted to draw everyone onto its home playing field, the free market. To this end, they established a high-level company task force, the Public Policy Study Group, to devise and promote a plan. The plan they were soon shopping around Congress was basically one in which the government could support farm income all it wanted, just so long as it didn't do so by raising commodity prices or unduly controlling production. Instead, the government should leave prices alone and make separate, direct payments to farmers during tough times. Cargill did not originate the direct payment concept; variations on the theme had been around for years. Cargill can, however, lay claim to playing a key role in making the idea politically acceptable.

In the early 1960s the government supported the price of corn and wheat with a program of loans. If a farmer agreed to set aside some land and thereby help control production, the government would guarantee that farmer a relatively high price for his or her crop. The guarantee took the form of a loan against the crop that need not be repaid. The government would first set a price it thought fair for farmers, then loan them that much for every bushel of grain farmers put into storage. If a farmer could later sell the grain at a price higher than the loan rate, he or she was free to do so and repay the government loan from the proceeds. Otherwise, the farmer could keep the money and the government would keep the grain. By calling the price support payment a loan, farmers were spared the indignity of receiving a direct welfare payment from the government. And since participation in the program was voluntary, farmer freedom was preserved.

The Cargill plan maintained the aspect of freedom but changed the payment method in a fundamental way. A farmer who agreed to set aside land would get a check called a "deficiency payment." That payment would be the difference between the world price and whatever price the government wanted farmers to have. For example, if corn was selling for $1.07 in the world market, and the government wanted farmers to have $1.25, the participating farmer would get a check for $0.18 per bushel directly from the government. An immediate and far-reaching consequence of the new program was that more grain would go straight to grain companies without ever being owned by the government. If U.S. corn was selling at or near the world price of $1.07, the government would not acquire any corn; all of it would be bought by grain companies at the prevailing price. Under the old program, the government would have acquired enough corn to cause the U.S. price to go up to $1.25.

Farmers at first howled about the direct payment provisions in the new plan. Faced with the onerous prospect of mandatory supply control, they started to come around. Then, in the wake of the wheat referendum disaster in 1963, the USDA bid farewell to the farm policies of Camelot and began pushing the emerging export market as the most likely salvation for farm income. Cochrane spent his last year in Washington helping put together a deficiency payment program for wheat. As for Cargill, the company's sales doubled in just five years. Sales by all U.S. farms combined were barely six times as large as the $2 billion grossed by Cargill in 1965.

Cochrane recovered slowly from his experience in Washington. Friends at the university did what they could to salvage the regents professorship disaster and soon had him appointed to the new position of dean of international programs, a change of pace that seemed to rekindle his spirits. The American Farm Economics Association awarded him their highest honor, appointment as a fellow. He also started consulting and brought in enough money to begin raising horses, a passion he had inherited from his father but could never afford. Most of all, he summoned the grim determination to do what he had always done so well. He began writing again. Within a year of leaving Washington, his delightful book *A City Man's Guide to the Farm Problem* was awaiting publication.

The liberal farm agenda was as important to him as ever. He remained the impassioned supporter of public programs for agriculture in a variety of forums and commissions, one of which had an unexpected ending. When President Johnson signed the Food and Agriculture Act of 1965, he first paid tribute to "the miracle of American agriculture" and praised the system that had "encouraged development of the family farm," then promised higher farm incomes, lower consumer costs, and surpluses vanishing in tides of new exports. Perhaps Congress was still a bit uncertain about the new deficiency payment approach they had adopted, for the act also established a National Advisory Commission on Food and Fiber. This select group of "Americans of broad experience and great talent" was to reexamine U.S. farm policy. Cochrane was among the thirty academics, farmers, and agribusiness leaders Johnson appointed. Sherwood Berg, by then dean of agriculture at the University of Minnesota, chaired the group.

The commission held a series of meetings around the country. One faction, led by the conservative economist D. Gale Johnson of the University of Chicago, developed a draft report favoring a return to the free market. Cochrane led the opposition by writing a report that defended government intervention. The final document was due and the last meeting was imminent. After long discussion concerning which version to adopt, Chairman Berg called for a vote. It was a tie. Berg, utterly exasperated, declared that he would flip a coin. The winning report would be submitted to the president on behalf of the entire commission. Shocked by Berg's impiety, Cochrane and D. Gale Johnson beat a hasty retreat to prepare a compromise document. Neither man could swallow the notion of the great Harvard vs. University of Chicago donnybrook being decided by something as mundane as a coin toss.

Cochrane found a better stage for his policy ideas in India. He had long been interested in the country, first as a history buff and later because India was a major recipient of American food aid shipments. Cochrane learned that food aid shipments were not as easily justified as they first appeared. On the one hand, starvation could be averted. On the other, constant imports of low-priced foods left many countries with little incentive to improve their own food production system. India, in particular, was often thought of as relying on U.S. food aid to stave off hunger while the country spent its main effort in developing a stronger urban core.

Cochrane first visited India for about a week in 1964 when he was still with the USDA. He was joined there by Secretary Freeman, and they toured both the poorest parts of the country and its most productive farming areas. They were trying to learn more about the degree to which U.S. food aid shipments were leading to economic development rather than economic dependence. They also were intrigued to see some of the new wheat and rice varieties that signaled the coming global "Green Revolution" in food production. In a second trip during that same year, he was more concerned with finding ways to improve the primitive way grain shipments from the United States were stored and handled when they reached India. Hand labor was used to bag grain in ship's holds for transport inland; barely 10 percent of India's 580,000 villages had access to suitable roads.

He renewed his acquaintance with Douglas Ensminger during his early trips to India. The two men had worked together in the old Bureau of Agricultural Economics. While Cochrane focused on domestic food policy, Ensminger had gone on to become the Ford Foundation's representative for India. Shortly after he left Washington, Cochrane was offered a position to be director of the agricultural program in India for USAID (U.S. Agency for International Development). He turned it down, partly because he did not want to relocate his growing family and partly because earlier trips abroad had left him with a sour impression of the USAID bureaucracy. Instead, Cochrane visited India every year from 1966 to 1970 as a consultant to the Ford Foundation for periods ranging from two weeks to three months.

In the mid-1960s poor weather in India raised the specter of widespread famine. Tragedy was averted with shipments of surplus U.S. wheat so large that, at long last, taxpayers got some relief from their grain storage headache. In 1963 the Commodity Credit Corporation owned over 1 billion bushels of wheat; in 1967 there was only 10 percent as much in public hands. Cochrane still talks with reverence of a vision of dozens of ships lined up and waiting to unload U.S. grain at an Indian port. This was the liberal dream of helping the poor and downtrodden at its best.

Ensminger, who was close to India's minister of food and agriculture, relied on Cochrane's economic advice for all aspects of the Ford Foundation agricultural program. Cochrane eventually prepared a report entitled *Food and Agricultural Policy* for India that, as much as anything he has written, lays out an example of agriculture functioning as a public utility. It must have been in some ways refreshing to work in an agricultural economy in which production need not be constantly restrained. He nonetheless was careful to acknowledge and plan for how the Green Revolution could bring India into the company of those countries struggling with surplus production.

In Cochrane's plan, the central government of India would operate a publicly held stabilization stock of millions of tons of grain. The stock "should be viewed as a giant balancing wheel in the food grain economy pouring grain supplies into commercial channels as needed to stabilize consumer price levels and withdrawing supplies at the farm level as needed to stabilize prices to farmers." He recognized that the stabilization stock operations would lose money more often than they made money, but the stock's effectiveness should be judged by how well it stabilized prices, not by whether it lost money in doing so. In fact, he saw the profit motive in then-current public efforts as being counterproductive.

The public, through its actions to stabilize prices, would thereby be setting prices. Since the prices set would have to be high enough to encourage farm production, the poor would not always be able to buy enough food. This was unacceptable. He therefore outlined a system in which the poorest citizens would be given ration cards with which they could obtain food. Exactly how to do this in a country of such widespread poverty was unclear; at one point he acknowledged that his idea could be impossibly expensive. Even so, he considered it to be among the most important provisions of his proposal.

Subsidized food imports from the United States were to be phased out as part of a plan to help India become more self-sufficient in food production. Such food imports, when necessary, would be used to subsidize wages for poor workers so that large-scale public works projects could be advanced. The roads and other infrastructure necessary to efficiently distribute food in India would thereby be improved. Cochrane left little doubt that he was not opposed to technology when it was clearly needed. As part of achieving a "modern scientific agriculture," he outlined extensive research and development programs for water management, plant disease prevention, and the breeding of higher-yielding varieties.

He also made it clear that he was not an opponent of private enterprise. In particular, he talked about how "more is required than improved government operations" if India was to adequately store and handle the massive amounts of food required to feed its population. He said that "the time has come to stop maligning the private trade" because "private marketing organizations with modern structures and equipment, with adequate credit, and with far-flung trading connections will be required to market the bulk of the crop." He recommended competition, from both private companies and farm cooperatives, as the best way to police the private trade "so that it does not gouge the farmer on one hand and the consumer on the other."

At this time Cochrane was, perhaps more than ever, in favor of policies to protect and encourage small farms. On the one hand, he recommended taxes on the earnings of larger, more productive farms. This could take the form of an income tax or, failing that, "a graduated land tax in which tax rates per acre increase with the number of acres operated." On the other hand, he laid out a three-part plan to support small farms. The management skills of smaller farmers would be improved, they would be given access to more credit, and cooperatives would be established to minimize the problems of scale, which seemed to always accompany technology in agriculture. It would take a massive effort to save the little farmer, but the effort could well be successful. The alternative was a system dominated by relatively few large farmers that would lead to social problems "beyond the comprehension of this writer."

His plan was widely discussed by officials in India and, in Cochrane's view, led to a much better understanding of agricultural economics on their part. The lasting effect of the plan is more debatable. There was a changing of the guard in the ministry and the food crisis passed on its own accord thanks to improved weather. Regardless of its practical adoption, *Food and Agricultural Policy for India* remains a fine example of liberal thinking as applied to food policy.

Cochrane settled back into the classroom during the 1970s and regained his status as one of the University of Minnesota's most sought-after teachers. His course in the economic history of U.S. agriculture was especially popular and led to what many consider to be his best book, *The Development of American Agriculture*. He was home more and as at peace with himself and his family as he had ever been. He remembers those years as ones of tranquility.

In the fall of 1969 he closed a deal on sixty acres of land north and east of St. Paul near the town of Stillwater. This was soon to be named Morningstar, a proper home for his family and a growing string of Morgan horses. He quickly built a barn, hay storage, a tack room, and an office on the place. Fortunately the office was heated, since it would be two more years before a house would be built. During those two years either he or his son Jamie would spend each night in the office caring for four horses and the foal Morningstar Magic.

Mary was not overly enthusiastic about leaving their comfortable home near the St. Paul campus for life in the country, a situation made worse by a fall she took from a runaway horse on the property. Their sons, too, began to tire of riding and driving horses. Undaunted, Cochrane pursued his ambitions into the world of horse shows. He quickly learned that the national circuit was no place for a college professor of modest means. The regional shows were a different story, however, and he took home his share of blue and purple ribbons from places as far away as Illinois and Colorado. "You go into the horse business for the love of horses, not to make money," he once told me as he listed the steady losses from his hobby.

In his personal papers Cochrane recounted a harrowing adventure with his horses. The scene is the fall of 1988 on a highway in California, where Will and Mary had originally retired. They were pulling a trailer with two of their mares, Mystique and Maisie, when suddenly:

> Mystique went crazy kicking the hell out of the trailer and Maisie who was next to her. I got stopped at a turn-out and ran back to the trailer. The partition had been kicked out and Maisie was laying flat on the floor of the trailer with Mystique on top of her still kicking up a storm. We unhooked the trailer; Mary drove off and found a phone and just by luck reached our veterinarian, who was a big man and a cowboy type of vet. He came right over and helped by literally pulling the two horses out of the trailer by their tails. What to do? We were at a pullout space on a busy highway with two half-crazy horses. I did not want to leave Mystique with some stranger who might sell her for meat. The vet had a large ranch of his own—so I asked the vet if he would put her down some place on his ranch. He said, "Yes." He got Mystique into his large stock trailer and took her to his nearby ranch. He had a man with a backhoe dig her a grave, he put her down painlessly and buried her under a large oak tree. I visited her grave several days later. That was the sad ending of my fine old mare, Mystique.

Loneliness and the California heat eventually persuaded Will and Mary that Minnesota winters weren't so bad after all. They moved back to Minnesota in 1989 and now live in a townhouse in Stillwater. Today, Cochrane still owns and cares for Jason, a stallion born of Mystique. Jason is boarded under the best of conditions not too far from the Cochrane home. As his personal writings show, the years have done nothing to diminish his great love for his horses: "Jason is now 25 years of age, and like his owner he is showing his age. He spends every good day in the pasture, which is good for his breathing. I go see him every Monday, Wednesday, and Friday morning, and take him carrots, which he likes and are good for him. I love the old boy dearly, and am doing my best to make his old age comfortable and pleasant."

During the 1970s Cochrane became increasingly disenchanted with farm policy. The treadmill began to take on a new look, first in *The City Man's Guide to the Farm Problem*, then in *The Development of American Agriculture*. In the original formulation farm profits were always driven down by falling prices.

But what would happen if prices were not allowed to fall? Supporting farm prices was, after all, a perennial mainstay of billion-dollar farm programs. In these cases farm profits would still stay low, but for different reasons. The price of land would rise, and landowners, not farmers, would get rich. To the extent farmers were also landowners, they would benefit as intended from farm price supports. Farmers who rented, however, saw their benefits go to absentee landlords; in Minnesota, farmland rental rates more than tripled during the 1970s. This was a fundamental departure from the world of *Farm Prices: Myth and Reality*. The word "land" is nowhere to be found in its long and thorough index. In the 1950s Cochrane thought consumers would be the inevitable beneficiaries of technology because it meant cheaper food. He later came to think that neither farmers nor consumers were clear winners. Landlords and, in his most recent thinking, agribusiness interests were well positioned to profit from farm technology.

Cochrane was also increasingly concerned over the number of farmers remaining in the United States. There were fewer than 3 million farms in 1970, less than half the number found in 1935. Each year during the 1970s another fifty thousand or so would disappear. As distressing as they were, declining farm numbers told only part of the story. Farm production was becoming concentrated among even fewer, larger farms. In *The Development of American Agriculture* he wrote that only 520,000 of the farms remaining in 1970 produced and sold 75 percent of the national farm product. On the other side of the scale, two out of three farmers in 1970 were part-timers whose off-farm income exceeded their income from farming.

Traditional farm programs were encouraging this move toward fewer and larger farms. Deficiency payments, like the price support programs they had replaced, were based on production. The more you produced, the more you were paid by the programs. Provisions in the programs to limit payments to farmers were easily sidestepped by smart farmers working with even smarter accountants. A more serious attempt to target payments toward smaller farmers was considered during the Carter administration in the late 1970s. President Carter lost to President Reagan in 1980, however, and the plan never got off the ground.

While many of his time did not perceive a problem with the move to larger and larger farms, Cochrane was not among them. He was concerned both for the individual trauma of those being forced out of farming and for the lasting negative effects their departure had on rural communities. He also worried that a time would come in which there would be so few farmers that they would control food as a monopoly and act against the public interest. This seems odd at first, since granting marketing orders was at the heart of his supply control concept. Those officially sanctioned monopolies, however, would have been closely regulated in his public utility concept. Monopolies brought about in the more conventional way would have no such public oversight. That commodity programs were helping, rather than correcting, this move toward larger farms eventually caused him to rethink the programs altogether. As he bluntly put it, we should "get government out of agriculture where it is helping one group of farmers do in another group."

In 1985 Professor Emeritus Cochrane was back before his colleagues at the American Agricultural Economics Association. There was quite the buzz over his title, and more so over the ideas presented. The title of the paper had his trademark flourish: "The Need to Rethink Agricultural Policy in General and to Perform Some Radical Surgery on Commodity Programs in Particular." The paper, in a nutshell, argued that "we should eliminate the price and income support features of the commodity programs as quickly as possible." The grand master of price supports was saying it was time to think of something else.

This was by no means to say that government should not be involved in agriculture. The public ought to be involved in providing a financial "safety net" for all farmers, in protecting the environment, and in feeding the poor both at home and abroad. He also offered specific ideas on how to encourage family farms. Dismayed that "there is no longer any such thing as the specific family farm," he argued for special management training, financial assistance, and credit sources for small and middle-sized farms. Research and education should be tailored more to their needs and less to those of larger farms. In addition, a national goal of getting one thousand young families into farming each year ought to be pursued.

When should the commodity programs be eliminated? "I, for one, will take the elimination of the commodity programs whenever we can get it, either all at once or gradually," he said. In a "rational world," however, the best solution would be to phase out the programs over a period of several years to give farmers time to adjust. This is essentially what happened in the 1996 "Freedom to Farm" legislation. Cochrane's suggested goal of increasing family farm numbers was nowhere to be found in the legislation.

Conversations with Willard Cochrane are never dull or predictable. One day you might be treated to a discourse on how rock music has so ruined today's young people that they can't even sit still, much less think for more than a few minutes at a time. His demonstration for me (in a public restaurant, by the way) of how they are "always bouncing, like this" is something I will not soon forget. Another time, you might hear an insightful and controversial analysis of how research in agricultural economics has lost its way because modern theoreticians can't see the world beyond their computer screens. At the very same lunch gathering, you can choke back laughter as a man in his eighties contemplates a visit to a female urologist and later be sobered by his latest thoughts on the implications of global warming. Mostly, though, our conversations have turned to what he now likes to call "agriculture and food system policy," rather than "farm policy," as a reminder that yesterday's remedies, regardless of past effectiveness, are unlikely to work in the coming years. The world has changed too much.

Cochrane sees the trend toward a more industrialized agriculture as continuing unabated unless new policies are put in place. Fewer and fewer independent farmers will survive as virtual monopolies set prices for inputs farmers use and for products they produce. The lack of competition at the agribusiness level will also mean that remaining farmers could be denied access to important supplies and markets. In addition, farms will continue to be plagued by unpredictable prices that have been of concern to Cochrane throughout his professional career. The global economy is inherently unstable and can lead to recessions of the type raging through the Midwest in the late 1990s. Current price problems give way to far more serious fears of production shortages as he looks ahead to weather patterns permanently disrupted by global warming.

Food quality and quantity will decline in the age of industrialized global agriculture. His assessment from many years ago that malnutrition is widespread and hunger is sometimes a problem in the United States remains; other countries will fare even worse. Rich and poor alike will be increasingly vulnerable to food-borne illness. Scares from bacteria in meats and chemical residues in fruits and vegetables grown in the United States are now so common that they are no longer front-page stories. The global economy brings us imported foods produced with substandard sanitation and treated with chemicals banned from use in the United States.

He is also concerned for the well-being of our physical and social environment. Farm chemicals, excess nutrient use, and widespread soil erosion have degraded water supplies across the country. He worries that biotechnology will challenge the environment in ways we have not anticipated. Threatening life forms, unlike any type of pollution we have seen, can be self-perpetuating and therefore virtually impossible to contain. The social environment, too, will decline as rural areas continue to lose farmers. In their place will be low-paid foreign and migrant workers who will need public support.

Problems such as these are beyond anything conventional farm programs were designed to handle. He calls Republican programs to get the government out of agriculture "pure nonsense." Even the late-1990s "mini-recession" was too much for the free market advocates. Money was again doled out, with the blessing of both major political parties, in much the same way it had been in the past. Production histories guided payment amounts; larger farmers gained more—much more—than their smaller neighbors. The traditional Democratic response of raising price supports is one Cochrane would have supported years ago. Now he sees such policies as subsidizing the largest farms at the expense of smaller farms, creating windfalls for landlords and bringing about problems for world trade. For these and other reasons, he sees no easy solutions for dealing with twenty-first-century industrial agriculture.

As I write this, Willard Cochrane believes more strongly than ever that this new industrialized agriculture is too important to be left to the vagaries of the free market. At the heart of his current thinking on policy is a program to help preserve the remaining family-sized farms. About 665,000 farms in the United States have sales between $20,000 and $250,000 per year. For these farms, he would guarantee their income with an annual payment ranging from $15,000 to $25,000 per year. The payment would be given regardless of crops produced or prices received. Another, even larger, group of farms have sales below $20,000 per year. He would not extend the program to these farms because they are part-timers whose principal source of income is nonfarm employment.

The payments to preserve family-sized farms are not intended to support prices and thereby advance the traditional goal of maintaining the incomes of all farmers. Rather, they have the effect of preserving independent farm decision making and providing enough people to maintain viable rural communities. The payments would not have the usual side-effect of encouraging large farms. Landlords, too, would have a more difficult time profiting from the payments because the ever-present pressure on farmers to expand and therefore bid up land values would be reduced.

The larger farms, while not eligible for the annual payment program, would participate with all farms in other forms of assistance. A Food Production Refinancing Agency would be established to provide loans for farms of any size during times of very low prices. Disaster programs would also be in place for use whenever natural disasters threatened farm production. Farmers of all sizes would also benefit from a grain reserve program designed to stabilize world prices and to act as the government's agent in providing world food aid.

Cochrane's goal of providing a nutritionally adequate, healthful supply of food to all Americans would be served in two important ways. First, the many public food assistance programs such as food stamps and school lunch programs should be maintained and enhanced. These enhancements would take the form of higher-quality food provisions and improved services for those in need of the programs. Second, the Food and Drug Administration would be greatly strengthened as a regulator of food quality. It would inspect all food consumed in the United States, from both imported and domestic sources, and have the authority to destroy all substandard food items.

Cochrane sees free and open competition as an essential part of a well-functioning food system. Competition in agriculture, while praised by nearly all, has often been left to a "fox guarding the chickens" strategy of relying on federal and state-level agricultural agencies. Cochrane would change this with the creation of a special unit in the Department of Justice with the sole charge of investigating monopolistic actions throughout the food production and distribution system. Prosecution would be more quickly forthcoming whenever lack of competition threatened any part of the food system. He would also have the president of the United States direct the Federal Trade Commission to investigate and prosecute those guilty of unfair business practices at any point in the food system.

To better protect the environment, he would establish a special unit of the Environmental Protection Agency to monitor and regulate the use of fertilizer, chemicals, and biotech farm inputs. The unit would have full authority to ban any of these items that harmed the environment. There would also be new legislation enacted to specifically target factory-style production and processing of poultry, beef, pork, and dairy products. The agency administering this program would have the authority to set and enforce standards in several key areas: the humane treatment of animals raised in confinement, the location of factory farms, the disposal of wastes from those farms, and the working conditions of people employed in factory farm operations.

Cochrane would also maintain the Conservation Reserve Program, a federal program that has since the mid-1980s retired millions of acres of erodible land from production. The program has operated in a way that rented fragile land from farmers or landlords on the condition that it not be farmed for ten years. Cochrane would like to see this land protected for much longer periods, sometimes up to one hundred years, and closer attention paid to make sure that only the most environmentally sensitive land was being retired. For land that could be combined into large ecological areas, he favors creation of a new agency to buy such land outright and combine it with other public lands to build new wildlife habitat regions.

Such a vast transformation of the American food system bears little resemblance to "get the government out of agriculture" strategies so dominant in the last several decades. When I asked him if this was what he meant by treating agriculture as a public utility, he said it came close but lacked one important element. He would not regulate food prices. Nonetheless, he made it very clear that now, as throughout his career, he views the food production and distribution system in the same way he sees our education and health care systems. Each must be in good working order at all times. Their products are not "market goods," they are basic human rights. For Cochrane, it is unthinkable that a private, unregulated system should be expected to guarantee human rights. That, above all, is the responsibility of public governance.

6

Heartland

I picked Will up at his home one summer morning in 1998. That I was thirty minutes early meant nothing, for he had somehow determined that we would never make it to the old farm in Iowa before nightfall. My assurances that midafternoon would be, as it always had been, a time of considerable daylight throughout the Midwest did little to calm him. We should have left at six; better yet, five. I grabbed his travel bag, boots, and briefcase from the driveway, bid a quick farewell to Mary, and we were off.

Time passes quickly when you are "Driving Mr. Willard." We turned our attention first to a studied analysis of how Tom Kelly could better manage the lackluster Twins, then to a long discussion of whether the North or the South had the superior leadership during the Civil War. Soon we were in Iowa, headed south on Interstate 35 past the large sign welcoming us to the Heartland. For over a hundred miles we saw nothing but corn, soybeans, and an occasional metal building in which unseen hogs or turkeys lived out their short lives. We saw not one single person working in any of the fields we passed, nor a single farm animal grazing on what had once been a great prairie of grass. Despondent farmers would soon mount two-hundred-thousand-dollar combines to begin gathering a near-record crop destined for sale at prices that, adjusted for inflation, ranked among the very lowest of the century.

The term *heartland* seems curious as you drive through these deserted fields. In the town where I live, they have taken to naming new housing developments after what used to be there. Monarch Meadows features unimaginative boxes fashioned from dull vinyl siding, all crowded on a nice little field where real monarch butterflies probably once flitted. On the other side of town, grotesque mini-mansions surrounded by monotonous Chem-Lawns trumpet the fruits of unbridled consumption from Mayflower Hill. Some of the old folks still remember gathering mayflowers there on Sunday picnics. So it seems this vast agricultural industrial park stretching in front of us was named: Heartland.

We drove to Des Moines, then west on Interstate 80 for another forty miles to the Stuart exit. It was a little after two in the afternoon (still daylight, I might add) when we crossed the railroad tracks and saw the boarded-up windows and faded Rock Island logo of a deserted railroad station. This was the very place where Will and his mother had arrived seventy-seven years before. Will's sharp mind, its memory sometimes clouded by advancing years, was crystal clear in recalling that day in 1921 when a horse and buggy had first taken him to the farm. It's curious how memory works. My father and my uncle Jack sometimes forget what they had for breakfast but remember the names, the hometowns, and the stories of each man who sailed with them in search of enemy submarines during World War II. My wife, too, sometimes walks through her grandmother's long-demolished house back in Morgantown, West Virginia, a room at a time, touching curtains here and a worn dining table there.

Cattle grazed on a hillside near the small church that marked our turn onto a dusty gravel road. Soon we reached the intersection where Will's Grandfather Chambers had driven his Model T at the precise time another car was approaching. The resulting fender bender, the first anyone could remember in those parts, was as unlikely then as it would be today. Not to worry, grandfather never liked cars anyway. Another turn amid land too rolling for monster machines, better suited for grass and cattle but often cultivated anyway, then we stopped and got out of the car. The farm lay before us.

The old two-story house, vacant for decades, was barely standing. It was the ashen color that old wood often turns when its last coat of paint has been stripped by the weather. The sun peered through gaping holes in the roof and broken windows, throwing shadows that made me think for a minute I saw someone moving inside. A few old oak trees pressing out from the foundation appeared to cradle and support the walls long after they should have crumbled. Between the road and the house a small field of knee-high grass, mostly foxtail, marked the comings and goings of the afternoon's prairie breezes.

I parked on the road in front of the property and we got out of the car. Will stood there, looking quite handsome in his plaid short-sleeved shirt and khaki pants. He extended his arm and crooked pointing finger. Starting at his left, he slowly and with great care put everything back in its proper place. First there was the river, once again full of carp and bullheads set to challenge the fishing skills of a young boy and his grandfather. A story came to mind of a time when the river was swollen from summer rains. A stranger and his family, his horse, and his buggy were washed away and never seen again.

He traced each field and planted them with his finger. No weeds, not on grandfather's farm. There was some corn, some wheat, some hay, some oats. The animals were back where they belonged, lucky to be tended by a man such as grandfather. As Will remembers it, he put their well-being before that of his own family. He pointed out where the windmill once pumped water and a small pile of boards behind the house that marked the creamery. A metal grain bin that someone had irreverently put beside the house faded away and grandmother's garden sprang back to its rightful place. The porch of the house rose from the crazy angle to which it had fallen and relatives sat talking after a hard day's work. Others were visiting in the parlor. Smells from the summer kitchen promised a wonderful meal.

As we stood there in the peaceful breeze of 1921, I lost track of time altogether. Will surprised me with stories of how often he had come here to visit. In one, he had camped in the yard with his boys decades ago. He told me, too, of the many times he had taken his mother here before her growing sadness at each visit and failing health confined her to California. His last visit had been five years before, during the week before his cousin Kenneth had died.

Will's mother is buried in the nearby town of Greenfield. Each deserted crossroad we passed through on our way there had once been a small community. Will for some reason recalled how his uncle Zene's son had walked to a store at one of those crossroads and bought the strychnine he used to end his life. Just outside of Greenfield we turned into a neatly kept cemetery where flowers and a few small flags were tucked among the headstones. The first area of plots to our left had a large memorial stone with the name CHAMBERS on it. Five smaller stones marked the graves of Will's mother, of Zene's son, of grandmother, of grandfather, and of Zene.

We drove back into town for a few minutes to visit a flower shop. After some discussion, Will and the proprietor agreed on a suitable arrangement for his mother's graveside. While we waited for the flowers, I noticed a gas station across the street and offered to get us both something cold to drink. Will thanked me and asked for a Green River. Then he laughed as he remembered that Green River soda hadn't been made for more years than he cared to remember. He settled for Sprite, a pale imitation at best. We sat there in the flower shop relaxing with our drinks, then delivered the flowers.

Will's cousin Louise was expecting us around five. We were right on schedule as we drove up to the new one-level house she had built a few years before. Louise, a few years younger than Will, greeted us and quickly apologized for her lack of energy now that she was enduring a third round of chemotherapy. That may have been, but my impression was more like that I have of my stepmother in southern Georgia, that of a woman strong enough to support an entire community if she had to. She was worried about Greenfield and the four farms whose pictures hung among the family portraits. She had inherited the farms as one relative after another had died. When she passed on, the farms would go to a daughter in another state who had shown little interest in either farming or Greenfield.

Professor Cochrane made a brief appearance and asked her thoughts on current farm policy. He was no match for Louise, however, and quickly retreated back to being Cousin Will as she told him in no uncertain terms that watching over her farms left little time for policy. We were a long way from Harvard, as close to the real Heartland as either of us was ever likely to get. I found myself thinking how nice it might have been if Will could have spent more time with Louise during his life. She could have told him not to worry so much about economic theory and the way it so quickly dismisses matters of the heart as being irrational or emotional. She could have reassured him that it was enough to hate what was happening to the home farm and that he would be foolish to feel otherwise.

The three of us joined Zene's daughter Donell and her husband for dinner at the Nodaway Cafe, then said our goodbyes. Will wanted to stop by the old farm one more time before we left. It was a short drive and we arrived just as the sun was setting on the fields behind the house. Light the color of fading fire filled the old place from a hundred breaks in its roofs and walls. Will patiently took it all in with the same look in his eyes that Louise had shown when she had hugged him goodbye for maybe the last time. He at long last turned back toward the car. With a Cochrane concoction of sadness, of bitterness, and of anger in his voice, he muttered, "Well, that's progress, I suppose."

Notes

1. Family Farms in Form but Not in Spirit

1. The quotes from Jefferson were taken from Gilbert Chinard, *Thomas Jefferson: The Apostle of Americanism* (Little, Brown, and Co., 1929), and Grant McConnell, *The Decline of Agrarian Democracy* (University of California Press, 1953). McConnell's book is also the source of the quote from President Theodore Roosevelt. The occasion for the quote was the initiation of the Commission on Country Life.

2. The full text of Henry A. Wallace's speech, "Thomas Jefferson: Farmer, Educator, and Democrat," can be found in E. E. Edwards, *Jefferson and Agriculture*, Agricultural History Series No. 7 (USDA, 1943).

3. Cochrane, *Farm Prices: Myth and Reality* (1958).

4. Cochrane, *The City Man's Guide to the Farm Problem* (1965).

5. *Farm Industry News*, September 1997.

6. *Farm Chemicals*, March 1997.

2. The Golden Age

1. D. J. Forrestal, *Faith, Hope, and $5,000: The Story of Monsanto* (Simon and Schuster, 1977).

3. The Treadmill

1. Quote taken from a headline in Monsanto's 1956 Annual Review.

2. Cochrane, "A Theoretical Scaffolding for Considering Governmental Pricing Policies in Agriculture" (1953).

3. Cochrane's views on the nature of supply and demand for food products are developed throughout his works. The presentation here is that given in *The Economics of American Agriculture*, chapter 24 (1951).

4. Professor Cochrane Goes to Washington

1. *Choices*, second quarter, 1996.
2. *Farm Chemicals*, January 1997.
3. *Monsanto Annual Report*, 1966.

5. An Unreconstructed Liberal

1. *Washington Post*, May 27, 1963.
2. Quoted in W. G. Broehl Jr., *Cargill: Going Global* (University Press of New England, 1998).

Selected Writings of Willard Cochrane, 1939–1997

"Organization and Practices of Financially Successful Montana Farms." Mimeo Circular Number 14. Montana State College. April 1939.

"A Price Policy for Agriculture, Consistent with Economic Progress, That Will Promote Adequate and More Stable Income from Farming." *Journal of Farm Economics*. November 1945.

High Level Food Consumption in the United States. Misc. Publication 581. USDA. December 1945.

What Peace Can Mean to American Farmers. Series of four USDA publications, prepared by a team including Cochrane:

Post-War Agriculture and Employment. Misc. Publication 562. USDA. May 1945.

Maintenance of Full Employment. Misc. Publication 570. USDA. July 1945.

Expansion of Foreign Trade. Misc. Publication 582. USDA. October 1945.

Agricultural Policy. Misc. Publication 589. USDA. December 1945.

"Farm Price Gyrations—An Aggregative Hypothesis." *Journal of Farm Economics*. May 1947.

"Family Budgets—Moving Picture." *The Review of Economic Statistics*. (Harvard University.) August 1947.

"Economics of Agriculture." In United Nations, Food and Agricultural Organization. *Report of the Mission to Siam*. September 1948.

An Analysis of Farm Price Behavior. Progress Report Number 50. Pennsylvania State College. May 1951.

"Output Responses of Farm Firms." *Journal of Farm Economics*. November 1951. (Coauthored with W. T. Butz.)

Economics of American Agriculture. First Edition. Prentice-Hall. 1951. (Coauthored with W. W. Wilcox.)

"A Theoretical Scaffolding for Considering Governmental Pricing Policies in Agriculture." *Journal of Farm Economics*. February 1953.

"Professor Schultz Discovers the Weather." *Journal of Farm Economics*. May 1953.

"The Nature of the Race between Food Supplies and Demand in the United States, 1951–75." *Journal of Farm Economics*. May 1953. (Coauthored with H. G. Lampe.)

An Economic Analysis of the Impact of Government Programs on the Potato Industry. Technical Bulletin 211. University of Minnesota. June 1954. (Coauthored with R. Gray and V. Sorenson.)

"The Case for Production Control." *The Metropolitan Milk Producers News*. December 1954.

"The Case for Production Controls." *Co-op Grain Quarterly*. St. Paul. April 1955.

The Economics of Consumption: Economic Decision Making in the Household. McGraw-Hill. 1956. (Coauthored with C. S. Bell.)

"Conceptualizing the Supply Relation in Agriculture." *Journal of Farm Economics*. December 1956.

"The Market as a Unit of Inquiry in Agricultural Economics Research." *Journal of Farm Economics*. February 1957.

"An Appraisal of Recent Changes in Agricultural Programs in the United States." *Journal of Farm Economics*. May 1957.

"On the Income Elasticity of Food Services." In *The Review of Economics and Statistics*. Harvard University. May 1957. (Coauthored with E. W. Bunkers.)

"The Case for Production Controls Restated." In *Policy for Commercial Agriculture*. Joint Economic Committee, 85th Congress. November 22, 1957.

Report of the Governor's Study Commission on Agriculture. State of Minnesota. 1958. (Cochrane chaired the commission and was director of research and editor of the report.)

Farm Prices: Myth and Reality. University of Minnesota Press. 1958.

"Some Additional Views on Demand and Supply." In *Agricultural Adjustment Problems in a Growing Economy*. Iowa State College Press. 1958.

Policies for Expanding the Demand for Farm Food Products in the United States: Part I History and Potentials. Technical Bulletin 231. University of Minnesota. April 1959. (Coauthored with J. Wetmore, M. Abel, and E. Learn.)

"Some Further Reflections on Supply Control." *Journal of Farm Economics*. November 1959.

"Farm Technology, Foreign Surplus Disposal and Domestic Supply Control." *Journal of Farm Economics*. December 1959.

"Public Law 480 and Related Programs." In *The Annals of the American Academy of Political and Social Science*. Volume 331. September 1960.

Selected Writings

Economics of American Agriculture. Second Edition. Prentice-Hall. 1960. (Coauthored with W. Wilcox.)

Food and Agriculture: A Program for the 1960's. USDA. March 1961. (Prepared and written under Cochrane's direction.)

Policies for Expanding the Demand for Farm Food Products in the United States: Part II Programs and Results. Technical Bulletin 238. University of Minnesota. April 1961. (Coauthored with M. Abel.)

"The Role of Economics and Statistics in the USDA." *Agricultural Economics Research.* July 1961.

"The World Food Budget: A Forward Look to 2000 and Beyond." *World Food Forum Proceedings.* USDA. May 1962.

"Long-Run Demand: A Concept, and Elasticity Demands for Meats." *Journal of Farm Economics.* August 1962. (Coauthored with W. G. Tomek.)

"Beliefs and Value Presuppositions Underlying Agricultural Policies and Programs." *Farm Goals in Conflict.* Iowa State University Press. 1963.

The City Man's Guide to the Farm Problem. University of Minnesota Press. 1965.

"Farm Land Prices and Farm Technological Advance." *Journal of Farm Economics.* May 1966. (Coauthored with R. W. Herdt.)

"Food and Agricultural Policy for India." Mimeo. The Ford Foundation. New Dehli, India. 1968.

The World Food Problem: A Guardedly Optimistic View. Thomas Y. Crowell. 1969.

"American Farm Policy in a Tumultuous World." *American Journal of Agricultural Economics.* December 1970.

Feast or Famine: The Uncertain World of Food and Agriculture and Its Policy Implications for the United States. National Planning Association. Washington DC. February 1974.

Agricultural Development Planning: Economic Concepts, Administrative Procedures, and Political Process. Praeger Publishers. 1974.

Economics of American Agriculture. Third Edition. Prentice-Hall. 1974. (Coauthored with W. Wilcox and R. Herdt.)

Reserve Stock Grain Models, the World and the United States, 1975–1985. Technical Bulletin 385. University of Minnesota. 1976. (Coauthored with Yigal Danin.)

"Economic Consequences of Federal Farm Commodity Programs." *Agricultural Economics Research.* April 1976. (Coauthored with F. J. Nelson.)

American Farm Policy, 1948–1973. University of Minnesota Press. 1976. (Coauthored with M. E. Ryan.)

"The Price of Farm Products in the Future." *The ANNALS*, the New Rural America. January 1977.

"The World Perspective." In *Rural Australia: The Other Nation.* Sydney, Australia: Hodder and Stoughton. 1977.

The Development of American Agriculture: A Historical Analysis. University of Minnesota Press. 1979.

"International Commodity Management as a Policy Problem for the United States: The Grains Case." In *The New International Economic Order: A U.S. Response.* New York University Press. 1979.

"Some Nonconformist Thoughts on Welfare Economics and Commodity Stabilization Policy." *American Journal of Agricultural Economics.* August 1980.

"The Economic Research Service: 22 Years Later." *American Economics Research.* April 1983.

Agricultural Economics at the University of Minnesota 1886–1979. Misc. Publication 21. University of Minnesota. 1983.

"The Need to Rethink Agricultural Policy in General and to Perform Some Radical Surgery on Commodity Programs in Particular." *American Journal of Agricultural Economics.* December 1985.

"The Long, Slow Slide into Economic Mediocracy." *Choices.* Third Quarter 1992. (Coauthored with H. Von Witzke.)

Reforming Farm Policy: Toward a National Agenda. Iowa State University Press. 1992. (Coauthored with C. F. Runge.)

The Development of American Agriculture: A Historical Analysis. Second Edition. University of Minnesota Press. 1993.

"The Troubled American Economy—An Institutional Policy Analysis." *Review of Agricultural Economics.* Spring/summer 1997.

"The Treadmill Revisited." *Land Economics.* October 1997. (Coauthored with R. A. Levins.)

"The New Deal and the Evolution of Farm Policy." *Southern Business and Economic Journal.* October 1997.

In the Our Sustainable Future series

Volume 1
Ogallala: Water for a Dry Land
John Opie

Volume 2
Building Soils for Better Crops: Organic Matter Management
Fred Magdoff

Volume 3
Agricultural Research Alternatives
William Lockeretz and Molly D. Anderson

Volume 4
Crop Improvement for Sustainable Agriculture
Edited by M. Brett Callaway and Charles A. Francis

Volume 5
Future Harvest: Pesticide-Free Farming
Jim Bender

Volume 6
*A Conspiracy of Optimism: Management of the
National Forests since World War Two*
Paul W. Hirt

Volume 7
Green Plans: Greenprint for Sustainability
Huey D. Johnson

Volume 8
*Making Nature, Shaping Culture: Plant
Biodiversity in Global Context*
Lawrence Busch, William B. Lacy, Jeffrey Burkhardt,
Douglas Hemken,
Jubel Moraga-Rojel, Timothy Koponen,
and José de Souza Silva

Volume 9
Economic Thresholds for Integrated Pest Management
Edited by Leon G. Higley and Larry P. Pedigo

Volume 10
Ecology and Economics of the Great Plains
Daniel S. Licht

Volume 11
Uphill against Water: The Great Dakota Water War
Peter Carrels

Volume 12
*Changing the Way America Farms: Knowledge and Community
in the Sustainable Agriculture Movement*
Neva Hassanein

Volume 13
Ogallala: Water for a Dry Land, second edition
John Opie

Volume 14
Willard Cochrane and the American Family Farm
Richard A. Levins

Archbishop Alemany Library
Dominican University
San Rafael, California 94901

HD
1771.5
.C63
L48
2000

Levins, Richard A.
Willard Cochrane and the
American family farm